Lecture Notes in Control and Information Sciences

Edited by M. Thoma and A. Wyner

Vol. 117: K.J. Hunt
Stochastic Optimal Control Theory
with Application in Self-Tuning Control
X, 308 pages, 1989.

Vol. 118: L. Dai
Singular Control Systems
IX, 332 pages, 1989

Vol. 119: T. Başar, P. Bernhard
Differential Games and Applications
VII, 201 pages, 1989

Vol. 120: L. Trave, A. Titli, A. M. Tarras
Large Scale Systems:
Decentralization, Structure Constraints
and Fixed Modes
XIV, 384 pages, 1989

Vol. 121: A. Blaquière (Editor)
Modeling and Control of Systems
in Engineering, Quantum Mechanics,
Economics and Biosciences
Proceedings of the Bellman Continuum
Workshop 1988, June 13–14, Sophia Antipolis, France
XXVI, 519 pages, 1989

Vol. 122: J. Descusse, M. Fliess, A. Isidori,
D. Leborgne (Eds.)
New Trends in Nonlinear Control Theory
Proceedings of an International
Conference on Nonlinear Systems,
Nantes, France, June 13–17, 1988
VIII, 528 pages, 1989

Vol. 123: C. W. de Silva, A. G. J. MacFarlane
Knowledge-Based Control with
Application to Robots
X, 196 pages, 1989

Vol. 124: A. A. Bahnasawi, M. S. Mahmoud
Control of Partially-Known Dynamical Systems
XI, 228 pages, 1989

Vol. 125: J. Simon (Ed.)
Control of Boundaries and Stabilization
Proceedings of the IFIP WG 7.2 Conference
Clermont Ferrand, France, June 20–23, 1988
IX, 266 pages, 1989

Vol. 126: N. Christopeit, K. Helmes
M. Kohlmann (Eds.)
Stochastic Differential Systems
Proceedings of the 4th Bad Honnef Conference
June 20–24, 1988
IX, 342 pages, 1989

Vol.127: C. Heij
Deterministic Identification
of Dynamical Systems
VI, 292 pages, 1989

Vol. 128: G. Einarsson, T. Ericson,
I. Ingemarsson, R. Johannesson,
K. Zigangirov, C.-E. Sundberg
Topics in Coding Theory
VII, 176 pages, 1989

Vol. 129: W. A.Porter, S. C. Kak (Eds.)
Advances in Communications and
Signal Processing
VI, 376 pages, 1989.

Vol. 130: W. A. Porter, S. C. Kak, J. L. Aravena (Eds.)
Advances in Computing and Control
VI, 367 pages, 1989

Vol. 131: S. M. Joshi
Control of Large Flexible Space Structures
IX, 196 pages, 1989.

Vol. 132: W.-Y. Ng
Interactive Multi-Objective Programming
as a Framework for Computer-Aided Control
System Design
XV, 182 pages, 1989.

Vol. 133: R. P. Leland
Stochastic Models for Laser Propagation
in Atmospheric Turbulence
VII, 145 pages, 1989.

Vol. 134: X. J. Zhang
Auxiliary Signal Design in Fault
Detection and Diagnosis
XII, 213 pages, 1989

Vol. 135: H. Nijmeijer, J. M. Schumacher (Eds.)
Three Decades of Mathematical System Theory
A Collection of Surveys at the Occasion of the
50th Birthday of Jan C. Willems
VI, 562 pages, 1989

Vol. 136: J. Zabczyk (Ed.)
Stochastic Systems and Optimization
Proceedings of the 6th IFIP WG 7.1
Working Conference,
Warsaw, Poland, September 12–16, 1988
VI, 374 pages. 1989

For information about Vols. 1–116 please contact your bookseller or Springer-Verlag

Lecture Notes
in Control and Information Sciences 181

Editors: M. Thoma and W. Wyner

C. R. Drane

Positioning Systems
A Unified Approach

Springer-Verlag Berlin Heidelberg GmbH

Author

Prof. Christopher R. Drane
School of Electrical Engineering
University of Technology, Sydney
P.O. Box 123
Broadway NSW 2007
Australia

ISBN 978-3-540-55850-7 ISBN 978-3-540-47298-8 (eBook)
DOI 10.1007/978-3-540-47298-8

Originally published by Springer-Verlag Berlin Heidelberg New York in 1992.

Typesetting: Camera ready by author

60/3020 5 4 3 2 1 0 Printed on acid-free paper

Dr. George Vorlicek
1948 - 1988

Acknowledgements

I am very grateful to Professor Trevor Cole from University of Sydney for providing sanctuary to undertake the initial stages of this work. I thank Professor Rod Belcher and Professor Warren Yates from the University of Technology for support during the latter stages of the study. Thanks also to Dr Godfrey Lucas, Mr Mark Johnson, Dr Hong Yan, Dr Nicholas Birrell, Professor Henning Harmuth, Mr Craig Scott, Mr John Drane, Dr Greg Searl, Dr Doug Gray, Mr Rick Jelliffe and the unknown reviewers for reading various versions of this manuscript. Ms Carol Gibson is responsible for much of the detailed LaTeXand many of the pictures to be found in this document.

Preface

This monograph had an unusual genesis. In the 1970s Professor Harry Messel had funded a research team at Sydney University to investigate novel techniques for tracking crocodiles. After developing a crocodile tracking system, the team turned their attention to tracking vehicles in urban areas. In the early 1980s, Dr Michael Yerbury, Dr George Vorlicek, Mr Jorg Suchau and I invented a prototype spread spectrum urban tracking system. By 1983 we had demonstrated very accurate tracking in urban areas. Over the next few years this created considerable commercial interest, both in Sydney and overseas. In 1987 I became the technical director of a company that was set up to develop a commercial version of the spread spectrum vehicle tracking system.

Another group in Sydney was working on a system with similar capabilities (lead by Dr Yerbury), and I was aware of several other such systems being developed at other places around the world. As well, the Global Positioning System was starting to come on-line.

It seemed to me that positioning systems were likely to move from the defence sector to become of great commercial interest over the next twenty years. The major area in which this would be likely to occur would be in the provision of up-to-date position information for people, land-based vehicles, aeroplanes and ships. This would coincide with the continued expansion of mobile communications systems. I believed that within twenty years, most mobile communications systems would also have a positioning capability.

Despite this likely scenario, it seemed that no-one was attempting to develop a unified approach to the overall specification and design of such systems. I saw such a unified approach providing a strong theoretical underpinning to these new positioning systems, as well as increasing understanding of more conventional systems such as radars, sonars and radio telescopes.

I believed this to be an excellent area for academic inquiry. Accordingly I spoke to research groups at a number of Universities with the aim of encouraging research in this area and had persuaded George Vorlicek to take up the challenge. I had even persuaded the financial backer of the company to fund University research of this nature.

All seemed set, but fate intervened. The financial backer ran out of money and Dr Vorlicek died suddenly in tragic circumstances. The company became insolvent and I became unemployed. I reconsidered the joys of University life and decided to resume my academic career.

Despite the failure of the company, I still believed that positioning systems would have a bright commercial future. Accordingly I set out to develop the theoretical framework which is described in this monograph. Initially I tried an approach based on Estimation theory, however I found that an information theoretic approach was much more rewarding. This was also a nice fit, as the most likely industry sector to commercially exploit positioning systems will be the communications industry.

Because the book is grounded in positioning systems rather than information theory, researchers trained in information theory may find my notation unorthodox. However, it is hoped that they can look beyond this. Rather than fully solving the many problems posed by the need to unify work on positioning systems, I have endeavoured to formulate the problems in such a form that researchers with expertise in information theory can become aware of the problems and perhaps solve them.

Despite the academic imperative of 'publish or perish', I found that the comprehensive nature of the work did not allow publication in a serial manner. I felt that all of the work would have to be published together in order for the technical community to properly assess its worth. This is why I have chosen the rather old-worldly method of first publishing my research in monograph form.

George Vorlicek looked forward to the day when vehicle tracking systems would bring large economic, safety and security benefits to the community. I hope this monograph can contribute to that vision.

Chris Drane
Foundation Professor of Computer Systems Engineering
University of Technology, Sydney

Table of Contents

1. Introduction

Positioning systems measure the location of one or more objects. Examples include Loran [32, 82], Omega [60, 113], Global Positioning System (GPS) [83], inertial guidance systems [68, 60], radars [109], sonars [118, 63] and urban vehicle tracking systems [48, 25, 24]. As well, most imaging systems inherently perform a positioning function. Examples of imaging systems include: magnetic resonance imaging, tomography, ultrasound [78], fluoroscopy, radiography [79], optical telescopes and radio telescopes.

Positioning systems have very wide areas of application. For example GPS will be applied to most forms of transport: general aviation [28], space navigation [74] and automobiles [21]. As well, various schemes for differential GPS will be applied to geodetic surveying [30, 75, 70, 115]. Because of the wide area of applicability of positioning and imaging systems, any general characterisation could be used in many areas of science and engineering.

The aim of this monograph is to establish a general method for the mathematical analysis of positioning systems and to demonstrate the utility of this characterisation. The basic approach uses information theory [103], differential geometry [124], rate distortion theory [8] and chaos [104], as well as integrating the more conventional methods of analysis based on estimation and detection theory [95, 107].

The first step is to derive a general formulation. This formulation is then applied to systems that use the propagation properties of waves to measure position. For example, systems that measure direction of a plane-wave arrival and the finite propagation time of sound or electromagnetic waves. This restriction to wave-based systems is not serious, as most important systems fall into this category e.g. radars, sonars, radio-telescopes, GPS and Omega.

A wave-based positioning system consists of one or more reference sites. The position of remote objects are measured relative to these sites. Each reference site may have a transmitter or a receiver or both. Each remote object may have a transmitter, a receiver, reflective properties or some combination of these. For example, in GPS each satellite transmitter is a reference site and a mobile vehicle will have a GPS receiver which picks up the signals from the satellites. In a simple radar system there is one reference site fitted with a transmitter/receiver and the targets scatter the radio frequency energy back to the reference site.

Positioning systems can be divided into two categories: *self-positioning* and *remote-positioning*. In self-positioning the remote objects measure where they are. In remote-positioning the system measures where remote objects are. GPS is a self-positioning system. A radar is a remote-positioning system. Of course an inherently self-positioning system can serve the purpose of a remote-positioning system if each remote object broadcasts its position using normal communication links. Similarly an inherently remote-positioning system can serve the purpose of a self-positioning system. Some systems may carry out a remote and a self-positioning function at the same time.

There has been a great deal of analysis of positioning systems, particularly radars [107]. Estimation and detection theory have been extensively applied to understanding the performance of single receivers [107, 95]. Estimation theory, especially in the form of Kalman filtering [55, 56] has been applied to improving the output measurements [106, 4]. Information Theory has been used to optimise a single radar receiver[129, 127, 128], to calculate ranging receiver acquisition times [26], energy management for surveillance radars [116], information content of Synthetic Aperture Radar images [36], a lower bound on data rate [133] and information sufficiency of radar signals [69].

None of these investigations takes a general, whole system approach to the analysis. Most of the work focuses on the performance of a single receiver which measures the position of a few objects. Analyses at the system level treat the measurements out of context to the system that made them. Historically this has been because most remote-positioning systems have only one reference site and in self-positioning systems it is only necessary to consider a single receiver at a time. As well, self-positioning systems only measure the position of one object, while most remote-positioning systems only measure a few hundred objects at a time.

Currently very large remote-positioning systems are being designed or built. These include Geostar [31], and terrestrial based urban vehicle tracking systems [48]. These systems share the characteristics of having to track large numbers of objects (of the order of one thousand to one million) and use multiple (of the order of ten) cooperating reception sites.

The proper design of systems like these requires a method of analysis that takes a whole-system approach, particularly with regard to system performance and system layout. The method of analysis presented here will be most useful for such large systems, but can be applied to simpler systems.

The extensive nature of this monograph should provide the basis to establish a theory of positioning systems. In order to be successful, any theory of positioning systems should satisfy the following requirements. It should:

- be applicable to a wide range of positioning systems,
- encapsulate a general notation which facilitates communications between users of different types of positioning systems,
- provide a definition of performance and practical methods of measuring the performance,
- establish an upper limit to the theoretical performance of a system,

- provide a way of determining the optimal configuration for a positioning system,
- be able to compare objectively the performance of two different types of positioning systems,
- provide a method of reasoning about classes of systems, and so allow general results to be developed,
- allow the direct comparison of positioning and communication systems,
- provide a method of determining the best measurement strategy for a system, including calculating the optimal resolution,
- establish a basis for optimising the process of selection of waveforms for wave-based positioning systems,
- provide a measure of how much information is being produced by the remote objects (or sources),
- specify the ambiguity of a positioning system and quantify the effect of ambiguity on system performance, and
- provide a way of specifying positioning systems which allows greater flexibility to the designer by not pre-judging technical issues.

This monograph outlines a formulation for describing positioning systems and establishes the feasibility of using this formulation to satisfy all the above requirements of a theory of positioning systems. More work needs to be done to establish a full theory, however it is believed that this monograph poses the remaining problems in such a way that specialists, particularly information theorists, will be able to solve the problems without an extensive knowledge of positioning systems.

The monograph is is divided into a number of chapters. Chapter 2 provides the overall formulation of the theory. Constraints on the capacity of positioning systems are derived in Chapter 3. Chapter 4 discusses methods of system optimisation and shows how it is possible to reason about classes of positioning systems. In Chapter 5 the process of coding for a positioning system is discussed while the process of decoding is analysed in Chapter 6. Chapter 7 places estimation techniques in the context of the positioning theory. The overall conclusions are presented in Chapter 8. To assist the reader, a glossary of commonly used symbols appears in Appendix A and there is a brief review of information theory in Appendix B.

2. Formulation

2.1 General Formulation

Consider[1] a positioning system making measurements on an n-dimensional space bounded by a domain of interest V_n. Suppose that there are m objects located in the domain, these objects will be labelled uniquely by the integers $1 \dots m$. A measurement of position on one of these objects will be in terms of the actual position $\boldsymbol{x}$ and the estimated position $\widehat{\boldsymbol{x}}$. Here $\boldsymbol{x}$ and $\widehat{\boldsymbol{x}}$ are n-vectors.

This can be seen as an information channel where the actual position is the input and the corrupted measured position is the output (see Fig. 2.1). The term channel is used in a general sense [123], including not only the spatial channel (e.g. the radio link) but the positional transducers and geometric effects.

Note that $\boldsymbol{x}$ may include phase space co-ordinates e.g. $\boldsymbol{x} = (x, \frac{dx}{dt})$. It is possible to send other information such as object shape, but such systems are part positioning system and part imaging systems. Here we will only consider systems which send phase space information. These will be referred to as *pure positioning systems.*

Each of the objects being measured by the positioning system can be represented by a stochastic vector process (the case of motion represented by a deterministic process is discussed in Sect. 5.1). Denote $\{\boldsymbol{\beta}_j(t)\}$ as the stochastic process representing the position of the jth object. Over a finite time period T the system operator will choose to make K measurements on the objects in the system, so defining a measurement vector. This vector can be written as

$$\boldsymbol{M}_T = (\gamma_1, \dots, \gamma_K)^{\mathrm{T}} \tag{2.1}$$

where γ_i is an ordered triplet (t_i, l_i, j_i) and the superscript T denotes transposition. The t_i is the start time of the measurement, l_i is the time taken to make the measurement and j_i is an integer indicating which object is being measured. For example if $\boldsymbol{M}_T = (\gamma_1, \gamma_2)^{\mathrm{T}}$ where $\gamma_1 = (3, 0.05, 7)^{\mathrm{T}}$ and $\gamma_1 = (3.05, 0.10, 25)^{\mathrm{T}}$ this means that at $t = 3$ seconds, the seventh was measured for 0.05 seconds and that $t = 3.05$ seconds the 25th object was measured for 0.10 seconds.

Assume the movement of the object is negligible during the period of the measurement. Then the measurement vector $\boldsymbol{M}_T$ defines a vector of positions $\{\boldsymbol{x}_1\}, \dots, \{\boldsymbol{x}_K\}$ with

[1] An early version of this formulation appears in Drane [22].

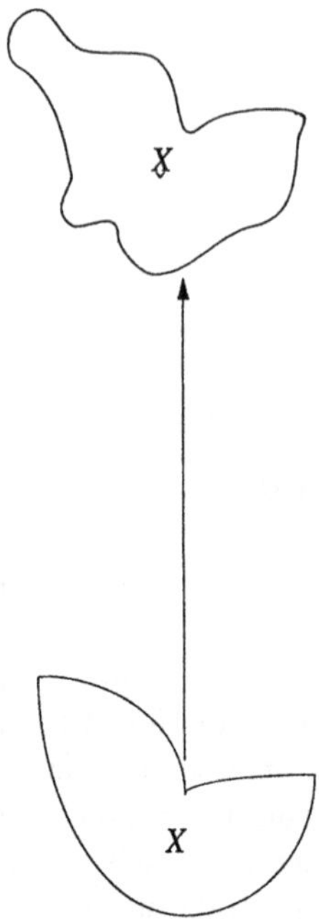

Fig. 2.1. Diagram of General Formulation

$$\boldsymbol{x}_i = \{\boldsymbol{\beta}_{j_i}(t_i)\}, \tag{2.2}$$

where $\{\cdot\}$ denotes a random variable.

The ith component of $\boldsymbol{x}_j$ will be denoted x_{ij}. For example, if a two-dimensional positioning system is using a x-y co-ordinate frame then $x_{2,5}$ would be the y co-ordinate of the fifth measurement.

Now the measurements yield a sequence $\{\hat{\boldsymbol{x}}_1\}, \ldots, \{\hat{\boldsymbol{x}}_n\}$. An important quantity is the entropy [103] of the vector of positions, which is given by

$$H(\{\boldsymbol{x}_1\}, \ldots, \{\boldsymbol{x}_K\}) = -\int \cdots \int_{V_n} \mathrm{d}\boldsymbol{x}_1 \ldots \mathrm{d}\boldsymbol{x}_K \, p_{\{X\}}(\boldsymbol{x}_1, \ldots, \boldsymbol{x}_K) \log p_{\{X\}}(\boldsymbol{x}_1, \ldots, \boldsymbol{x}_K) \tag{2.3}$$

or, using an obvious vector notation,

$$H(\{\boldsymbol{X}\}) = -\int_{V_n} \mathrm{d}\boldsymbol{X} \, p_{\{X\}}(\boldsymbol{X}) \log(p_{\{X\}}(\boldsymbol{X})) \tag{2.4}$$

where $\boldsymbol{X}$ is given by

$$\boldsymbol{X} = (\boldsymbol{x}_1, \ldots, \boldsymbol{x}_K)^{\mathrm{T}} \tag{2.5}$$

and $p_{\{X\}}(\boldsymbol{X})$ is the *a priori* probability density function (p.d.f.) of $\boldsymbol{X}$. The p.d.f. $p_{\{X\}}$ will be referred to as the *positional p.d.f.*

The reader will by this time appreciate that the notation has become cluttered. This is due to the generality of the formulation. Following [100], we will make a simplification by omitting the $\{\cdot\}$ notation for a random variable. The text will be used to clarify any possible ambiguity. As well, the subscript on the p.d.f. will be dropped unless there is a possibility of confusion.

The average mutual information [103] $I(\boldsymbol{X};\widehat{\boldsymbol{X}})$ represents the amount of information transfer between the input of the positioning system and the output. The average mutual information is defined as

$$I(\boldsymbol{X};\widehat{\boldsymbol{X}}) = \int_{V_n} \mathrm{d}\boldsymbol{X} \int_{V_n} \mathrm{d}\widehat{\boldsymbol{X}}\, p(\boldsymbol{X},\widehat{\boldsymbol{X}}) \log\left(\frac{p(\boldsymbol{X},\widehat{\boldsymbol{X}})}{p_X(\boldsymbol{X})p_{\widehat{X}}(\widehat{\boldsymbol{X}})}\right) \tag{2.6}$$

where $p_{X,Y}(\boldsymbol{X},\widehat{\boldsymbol{X}})$ is the joint p.d.f. of $\boldsymbol{X}$ and $\widehat{\boldsymbol{X}}$ and $p_{\widehat{X}}(\widehat{\boldsymbol{X}})$ is the p.d.f. of $\widehat{\boldsymbol{X}}$. The vector $\widehat{\boldsymbol{X}}$ is defined by

$$\widehat{\boldsymbol{X}} = (\widehat{\boldsymbol{x}}_1,\ldots,\widehat{\boldsymbol{x}}_K)^{\mathrm{T}}. \tag{2.7}$$

For a system that is making a number of measurements over a period of time, a useful parameter is the system *information performance*. If a system takes T seconds to make K measurements, the information performance in that time, P_T, is defined as

$$P_T = \frac{I(\boldsymbol{X};\widehat{\boldsymbol{X}})}{T}. \tag{2.8}$$

Later in this monograph there will be cases where the average mutual information will be a continuous function of time. In these cases it is possible to define the *instantaneous information performance* as

$$i(t) = \frac{\mathrm{d}I}{\mathrm{d}t}. \tag{2.9}$$

If the logarithm is to the base 2 in (2.6) then the entropy and average mutual information will be in units of bits and both the information performance and the instantaneous information performance will be in bits per second. It is suggested that this convention be adopted, emphasizing that positioning systems are a form of communications system, but for mathematical convenience, most algebraic manipulations will be performed using natural logarithms, denoted "log".

Example 2.1

Consider a one-dimensional system that makes K measurements of a single object. The time interval between measurements is t_m, and this interval is large enough that the measurements are independent. Each of the x_i are gaussian random variables with variance σ^2. The $\widehat{x}_i$ are given by $\widehat{x}_i = x_i + z_i$ where the z_i are gaussian random variables with variance τ^2.

Then from Fano [29, page 146], the average mutual information for a single measurement is given by

$$I(\boldsymbol{x};\widehat{\boldsymbol{x}}) = \frac{1}{2}\log\left(1+\frac{\sigma^2}{\tau^2}\right), \tag{2.10}$$

so that for K independent measurements, we have

$$I(\boldsymbol{X};\widehat{\boldsymbol{X}}) = \frac{K}{2}\log\left(1+\frac{\sigma^2}{\tau^2}\right). \tag{2.11}$$

Substituting this expression into (2.8) gives

$$P = \frac{1}{2t_m}\log\left(1+\frac{\sigma^2}{\tau^2}\right). \tag{2.12}$$

For instance, if σ is 5 120 metres, τ is ten metres and t_m is 10 metres, then the information performance will be 900 bits/s. The performance in this case is a logarithmic function of accuracy. This accords with intuition. For most positioning systems there is a strong decreasing marginal utility of increased accuracy. For example an ambulance dispatch system can gain greatly by knowing where an ambulance is to within 1 kilometre. It performs slightly better if it knows the position to within 100 metres, but it will perform only marginally better if the position is known to 1 metre. □

Of course it is possible to choose other performance functions rather than P_T, although such performance functions are likely to be application specific. As well, the use of $I(\boldsymbol{X};\widehat{\boldsymbol{X}})$ allows the mathematical apparatus of information theory to be applied to positioning systems.

It will be strongly argued in this book that the information performance is an important parameter for the characterisation of positioning systems. However it is not the only important parameter and indeed many other parameters are discussed in this monograph. Moreover, any realistic system appraisal cannot be made purely on the basis of information performance or even a combination of the other technical parameters, any more than the conventional communications systems is designed purely on the basis of bits/s comparisons. Ultimately cost, quality and timescale considerations must apply, and the rigourous systems engineer will evaluate a system effectiveness function [2]. Nevertheless, information performance is a very important parameter for gaining greater insight into a positioning system's performance and for making key technical decisions.

A common indicator of performance is the system accuracy [109] which is the root mean square error (r.m.s.) of the positioning system. It can be defined as

$$A_w = \sqrt{\frac{\mathcal{E}\{(\boldsymbol{X}-\widehat{\boldsymbol{X}})^{\mathrm{T}}(\boldsymbol{X}-\widehat{\boldsymbol{X}})\}}{K}} \tag{2.13}$$

where $\mathcal{E}\{\cdot\}$ denotes the expectation. In general, accuracy is a useful parameter for self-positioning systems whereas both information performance and accuracy (amongst other parameters) are needed to analyse remote-positioning systems.

One difference between a normal communications system and a positioning system is that in a typical communications system, small deterministic time delays in the delivery of the information are not important. The situation is quite different in a positioning system, as often even a few seconds delay can

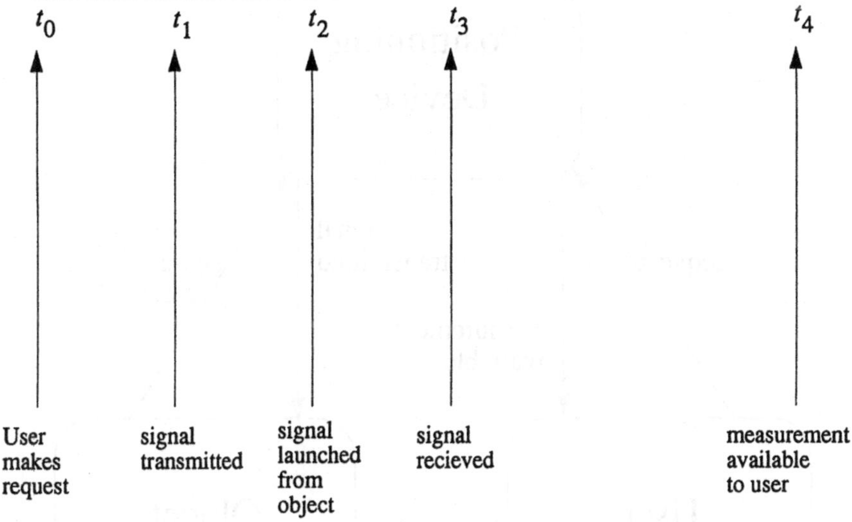

Fig. 2.2. Timing of Major Positioning System Events

make a position measurement useless. For example, in a fighter plane radar, a few seconds delay can cause an error of a kilometre or more. There are cases where this time delay will not be important, for example the data from a police surveillance tracking system may be used many years later in a court case. However on the whole, the amount of time delay is an important design issue for most positioning systems.

The time delay inherent in a system can be defined more precisely by considering Fig. 2.2. At time t_0 the user (which could be a human or a computer) makes a request to know the whereabouts of an object. At time t_1 the system broadcasts an interrogatory signal. For example, a radar might send out a pulse of Radio Frequency (RF) energy. At time t_2 the object reflects the signal or transmits a new signal. At time t_3 the signal is received by the positioning device and after processing, the measurement is available to the user at time t_4. The relationship between these various events is shown in Fig. 2.3. We can now make the following definitions for a particular measurement.

User Latency is the overall system time delay and is equal to $t_4 - t_0$. It is a measure of how long the user has to wait to get an answer.

Latency is a measure of how out of date the measurement is once it is presented to the user. It is equal to $t_4 - t_2$.

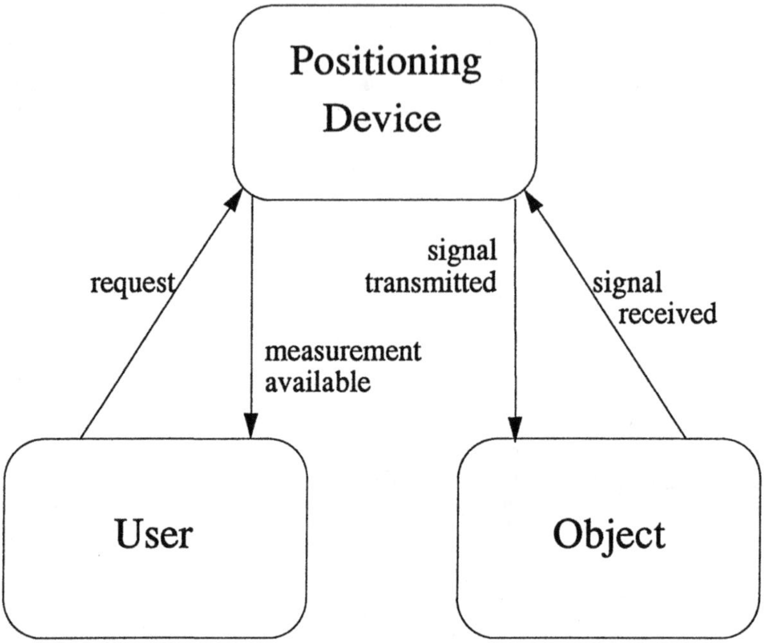

Fig. 2.3. Relationship of Major Positioning System Events

Physical Latency is the irreducible delay introduced by physical propagation and is equal to $t_3 - t_2$. In self positioning systems such as GPS this can be zero. In wave-based remote positioning systems this must be greater than zero. If there is more than one reference site being used to make the measurement, then t_3 is the time when the signal is received at the last site.

The physical latency will be a random variable because the distance to the reference sites will be a random variable. The mode of processing may also introduce extra variability. Each of these delays could be averaged over many measurements to yield an average user latency, latency and physical latency.

Because the delays are random variables, and we are normally trying to estimate not only a position, but also the time we are at a position, it would be possible to account for the extra information that the time estimate involves. However, it is normally possible, *a posteriori*, to estimate the time to a high degree of accuracy so it is not worth the extra complexity to include time in the information calculations.

It may be thought that because a large latency can make a position measurement useless, latency involves a loss of information. However this is a confusion between meaning and the formal definition of information [103]. For example, two messages can contain the same number of bits and hence the same amount of information, but one could be a dummy message whereas the other could be a Mayday call. Similarly two position measurements may involve the same amount of information transfer, but one could be so late as to be useless.

This section has defined information performance, accuracy and latency. A number of other parameters of positioning systems will be defined in subsequent chapters.

2.2 Wave-Based Systems

The formulation in the previous section is quite general, but it is difficult to make useful analytic progress with such a general formulation. In addition, the effects of the spatial channel, positional transducers, and geometrical factors are combined into a lumped transmission channel. This section applies the formulation for the important class of wave-based systems, which allows separation of the various factors.

Before proceeding further, it is worthwhile giving a few examples to clarify the concept of a wave-based system. Loran, Omega, sonars and radars are all wave-based systems, because the position information is coded directly onto the wave due to the physical properties of the wave propagation. Loran uses time delay, Omega measures phase shifts, sonars can use time delay or direction of wave-front arrival and radars use similar techniques to sonars. A mechanical gyroscope is not a wave-based system because no wave physics is involved. A compass/odometer system on a vehicle which then uses a radio link to transmit back the position data is not a wave-based system, because, although radio waves are used, the positioning information is not coded onto the wave due to the physical properties of the wave propagation, rather normal data modulation is used.

In wave-based systems, the user seeks to relate the position of the objects to an absolute, real-world cartesian measurement frame. Some systems may not use cartesian co-ordinates (e.g. some use latitude, longitude and height), but this can still be related to some absolute cartesian frame.

The actual measurements will be made in a natural co-ordinate frame determined by the nature of the positioning system. Initially this will be in terms of the measured characteristics of the wave propagation. For instance a radar might make measurements in terms of time delay and angle of arrival. This will often be transformed by an invertible transformation to a more useful form. For example most radars work in a polar (range-azimuth) or spherical (range-azimuth-elevation) co-ordinate system; GPS uses a pseudorange-pseudorange-pseudorange-pseudorange co-ordinate system.

The two most important features about this natural co-ordinate frame are that it is relative to the positioning system and that it introduces a distortion or even ambiguity. For example, a polar co-ordinate frame alters cartesian volume elements and so distorts the positional p.d.f. (as seen in polar co-ordinates). For most systems, common practice establishes the natural co-ordinate system. For some systems, an arbitrary decision has to be made. A method of classi-

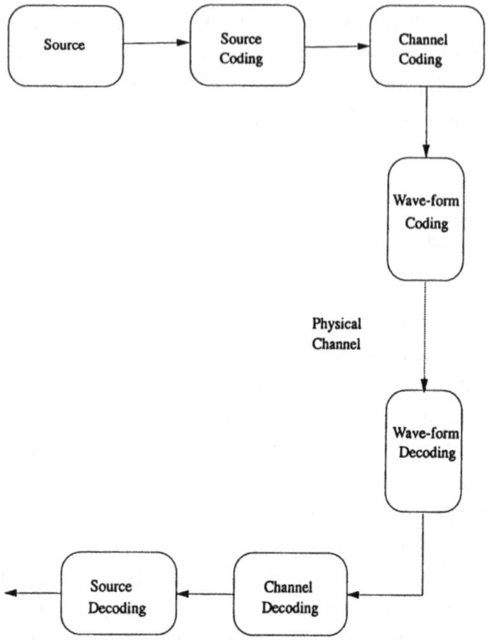

Fig. 2.4. Block Diagram of Conventional Communications System

fying systems is discussed in Appendix F. In any case, as long as a consistent transformation is made between the raw measurements and the chosen relative frame, the actual analysis is not affected (see Proposition 4.6 in Sect. 4.5).

Call the absolute cartesian measurement frame the *World frame* ($\mathcal{W}$). Let the positioning system's relative co-ordinate frame be called the *Communications frame* ($\mathcal{C}$). Normally, measurements made in the communications frame are transformed back to the world frame before being used.

The model of wave-based positioning system presented here is based on the model of a typical communications system. Accordingly we will first describe the model of a conventional communications system with reference to Fig. 2.4.

Here the *source* will produce the messages that need to be communicated. *Source coding* is the process of representing these messages in some sort of alphabet, for example Huffman coding. *Channel coding* is the task of efficient coding to account for channel properties, an example is Reed-Solomon coding. In order to transmit the signals through a physical medium, it is necessary to perform *waveform coding* or modulation. The signal can then be sent over the *physical channel* with a finite bandwidth and signal-to-noise ratio. At the receiver the signal must be demodulated (*waveform decoding*). Then the receiver must perform *channel decoding* and *source decoding.*

This model applies to *indirect positioning systems.* An indirect remote positioning system would operate by the remote object working out where it is using a wave-based system, and then transmitting the position information

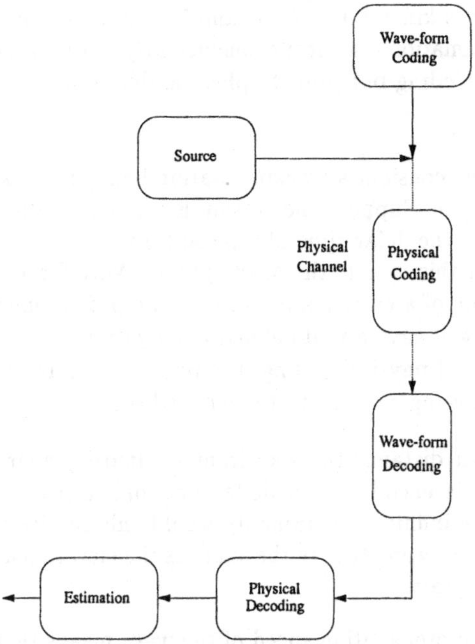

Fig. 2.5. Block Diagram of Wave-Based System

over a conventional communications channel. An indirect self positioning system works out the location of a remote object, then sends this information to the remote object using a conventional communications link. In either case the performance of the communications link could be described by the model shown in Fig. 2.4.

However this model does not apply to direct wave-based systems. For such systems, the model shown in Fig. 2.5 is advocated. Here the *source* is the moving object. The *waveform coding* is the process of the moving object emitting a signal modulated with a suitable waveform. Unlike the conventional communications model, this is done without reference to the message being sent by the source, although the chosen waveform can affect system performance by causing trade-offs between various parameters e.g. between range and range-rate. The signal is then transmitted over the physical medium, during which time *physical coding* occurs. It is during the process of physical coding that the position information is transformed from the world to the communications frame. This coding is due to the physical nature of the wave propagation and the overall configuration of the transmitters/receivers.

At the receiver, the signal must be demodulated, in the process called *waveform decoding*. This provides positional information expressed in the communications frame. *Physical decoding* is simply transforming this information to the world frame. Finally, a series of measurements can be combined, using *estimation* to form the final estimates. This process would normally use knowledge of the moving object's kinematics. An example is Kalman filtering. Note that in some systems estimation can occur concurrently with waveform decoding, or after waveform decoding but prior to physical decoding.

Example 2.2

As an example, consider a tracking station being used to measure the location of a space probe. Suppose the station is a polar positioning system which measures range by the delay time of a pulse train.

The source in this case is the space probe. Wave-form coding will consist of the broadcasting of a carrier wave which is amplitude modulated by a pulse train. The signal will have a well-defined polarization.

In the process of physical coding, the physical nature of wave-propagation will cause the following changes to the emitted signal.

Angle: At a great distance the wavefront is almost planar so that the angle information is encoded as a delta function, in other words a receiving antenna with infinite directionality would only receive the signal when its pointing angles were exactly the same as the elevation and azimuth angles of the space craft.

Range: The finite propagation speed of the wave means that the pulse will be received at a time different from when it is transmitted. In this way the range of the space craft is encoded as the time shift of a pulse. To simplify this discussion, assume that the space craft has a highly accurate clock, so that only a one-way transmission of the pulse is necessary.

At the receiver the signal has to be demodulated. This consists of cross-correlating the received pulse train with an internally held replica in order to estimate the time-shift. As well a directional antenna is used to work out the arrival angles (azimuth and elevation). These two operations correspond to waveform decoding. Next the azimuth-range information is converted to a suitable co-ordinate system, by the process of physical decoding. Finally a series of position measurements are combined using a suitable filter to produce the final estimates. □

Figure 2.5 gives a qualitative description of our model of wave-based positioning systems. Now let us develop the mathematical formulation of the model.

As before the source is described as a stochastic vector process which is operated on by a measurement vector to yield the vector of positions $\boldsymbol{X}$.

Let the mapping from the world frame to the communications frame be given by (see Fig. 2.6)

$$\boldsymbol{\xi} = \boldsymbol{g}(\boldsymbol{x}) \tag{2.14}$$

where $\boldsymbol{g}$ is an n-component vector function and $\boldsymbol{\xi}$ is the position in the communications frame of a point $\boldsymbol{x}$ in the world frame. The function $\boldsymbol{g}$ describes the geometrical factors of the system and so the process of physical coding.

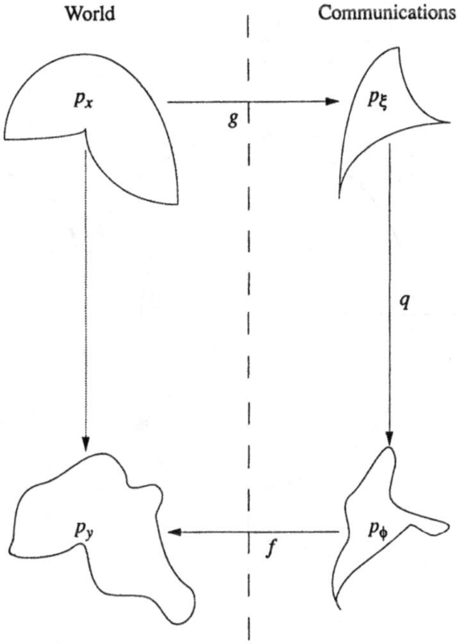

Fig. 2.6. Diagram of Wave-based Transformations

In order to send the position information from one point to another, the position information $\boldsymbol{\xi}$ must be encoded onto a spatial communications channel. This will corrupt the position information with noise and systematic offset. As well the positional sensors have to estimate position. The combination of spatial channel and sensors is referred to as the physical channel. An example of a physical channel is a pulsed ranging radar. Here the position of the object is encoded as the epoch of a pulse. At the receiver the epoch of the pulse is estimated in order to measure the object position.

Our treatment assumes that the estimated position in the communications frame results from the action of a random transformation [87], $\boldsymbol{q}$. This transformation characterises the physical communication channel. Accordingly the estimated position $\boldsymbol{\phi}$, as seen in the communications frame, is given by

$$\boldsymbol{\phi} = \boldsymbol{q}(\boldsymbol{\xi}). \tag{2.15}$$

Note that measuring $\boldsymbol{\phi}$ entails waveform decoding. Sometimes waveform decoding will use prior measurements, in which case $\boldsymbol{\Phi}$ will be a random transformation of $\boldsymbol{\Xi}$ where $\boldsymbol{\Phi} = (\boldsymbol{\phi}_1, \ldots, \boldsymbol{\phi}_K)^{\mathrm{T}}$ and $\boldsymbol{\Xi} = (\boldsymbol{\xi}_1, \ldots, \boldsymbol{\xi}_K)^{\mathrm{T}}$.

Next, the position as estimated in the communications frame will normally be transformed back to the original world frame, by the process of physical decoding. This can be done by a mapping

$$\boldsymbol{y} = \boldsymbol{f}(\boldsymbol{\phi}) \tag{2.16}$$

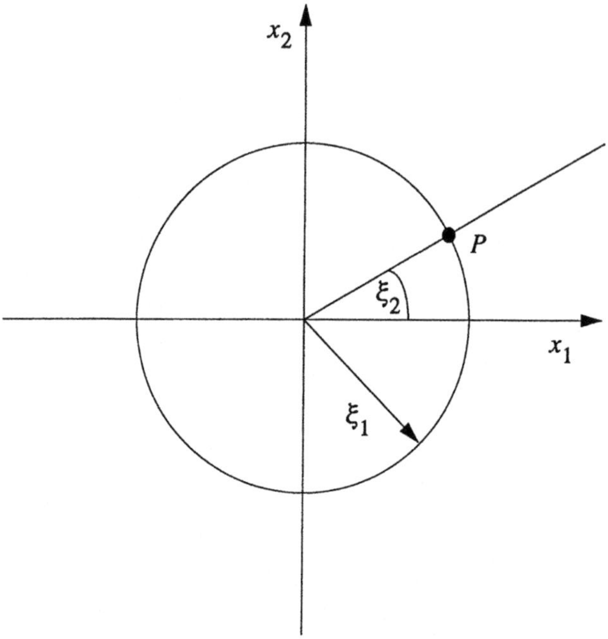

Fig. 2.7. Diagram of Polar System

where $\boldsymbol{f}$ is a n-component vector function. Equations (2.14), (2.15) and (2.16) define a mapping $\mathcal{W} \longrightarrow \mathcal{C} \longrightarrow \mathcal{W}$.

Finally an estimate can be made of the position, using knowledge of the object kinematics. This process is called estimation and is defined by

$$\widehat{\boldsymbol{X}} = \boldsymbol{F}(\boldsymbol{Y}). \tag{2.17}$$

The nature of these transformations will be clarified by the following example.

Example 2.3 Consider the polar positioning system discussed in Example 2.2. Assume zero-mean, additive measurement error $\boldsymbol{\zeta}$. The function $\boldsymbol{g}$ will be defined by (see Fig. 2.7)

$$\xi_1 = \sqrt{x_2^2 + x_1^2} \tag{2.18}$$

and

$$\xi_2 = \tan_2^{-1}(x_2, x_1). \tag{2.19}$$

The physical communications channel transformation $\boldsymbol{q}$ is defined by

$$\phi_i = \xi_i + \zeta_i \ ; \ i = 1, 2. \tag{2.20}$$

The transformation $\boldsymbol{f}$ is defined by

$$y_1 = \phi_1 \cos(\phi_2) \tag{2.21}$$

and

$$y_2 = \phi_1 \sin(\phi_2). \tag{2.22}$$

If the object is known to be stationary, a simple estimator $\boldsymbol{F}$ would be

$$\widehat{x_{ik}} = \frac{1}{k}\sum_{r=1}^{k} y_{ir} \quad ; \text{i=1,2; k=1,K.} \tag{2.23}$$

□

In summary, using these definitions we are able to describe our formulation of a wave-based positioning system precisely. Such a system consists of the source statistics ($\boldsymbol{\beta}_i$), the actual positioning device ($\boldsymbol{g}, \boldsymbol{q}, \boldsymbol{f}$) and the estimator $\boldsymbol{F}$. The function $\boldsymbol{g}$ represents the geometrical properties of the positioning device. The function $\boldsymbol{q}$ represents the properties of the physical communication channel, including how efficiently data are encoded on the channel. The function $\boldsymbol{f}$ will normally be chosen such that $\boldsymbol{f}(\boldsymbol{q}(\boldsymbol{g}(\boldsymbol{x})))$ is an unbiased estimator. The system operator will use such a system over a period T by defining a measurement vector $\boldsymbol{M_T}$ and by using the best possible estimator $\boldsymbol{F}$. More generally, a wave-based positioning system can be seen as performing a series of transformations:

$$\boldsymbol{X} \rightarrow \boldsymbol{\Xi} \rightarrow \boldsymbol{\Phi} \rightarrow \boldsymbol{Y} \rightarrow \widehat{\boldsymbol{X}}. \tag{2.24}$$

The formulation for wave-based systems is very useful as it allows a clear separation of the issues involved with source statistics, geometrical factors, physical transmission channel and measurement strategy.

3. System Capacity

Chapter 2 showed how a positioning system can be characterised in terms of its information performance. System designers can establish the capacity of a positioning system by calculating limits on the information performance. In Sect. 3.1, a useful proposition will be derived which will be used in Sect. 3.2 to establish an upper limit to performance. Then in Sect. 3.3 it will be shown that the performance measured in different reference frames will be the invariant for certain sorts of positioning systems.

3.1 Generalised Data Processing Theorem

A positioning system can be thought of as a generalised data processing system which operates on a number of random vectors (e.g. $\boldsymbol{X}$ and $\boldsymbol{Y}$).

If these vectors are discrete, scalar, random variables there is a theorem called the data processing theorem [53, 98] which sets limits on the information transfer between the various processes. The aim of the next proposition is to generalise this theorem to the case of continuous vector random variables.

Proposition 3.1 *Suppose $\boldsymbol{S}_1$,$\boldsymbol{S}_2$ and $\boldsymbol{S}_3$ are continuous, vector, random variables (i.e. like $\boldsymbol{X}$) and that they form a first order Markov chain. Then*

$$I(\boldsymbol{S}_1; \boldsymbol{S}_3) \leq I(\boldsymbol{S}_2; \boldsymbol{S}_3) \tag{3.1}$$

and

$$I(\boldsymbol{S}_1; \boldsymbol{S}_3) \leq I(\boldsymbol{S}_1; \boldsymbol{S}_2). \tag{3.2}$$

Proof

Consider $I(\boldsymbol{S}_1\boldsymbol{S}_2; \boldsymbol{S}_3)$ which we will define as

$$I(\boldsymbol{S}_1\boldsymbol{S}_2; \boldsymbol{S}_3) \equiv \int_{-\infty}^{\infty}\int_{-\infty}^{\infty}\int_{-\infty}^{\infty} \mathrm{d}\boldsymbol{S}_1 \, \mathrm{d}\boldsymbol{S}_2 \, \mathrm{d}\boldsymbol{S}_3 \, p(\boldsymbol{S}_1, \boldsymbol{S}_2, \boldsymbol{S}_3) \log \left[\frac{p(\boldsymbol{S}_1, \boldsymbol{S}_2, \boldsymbol{S}_3)}{p(\boldsymbol{S}_1, \boldsymbol{S}_2)p(\boldsymbol{S}_3)}\right] \tag{3.3}$$

where the limits of integration are suitably chosen.

Accordingly

$$I(S_1S_2;S_3) = \int_{-\infty}^{\infty}\int_{-\infty}^{\infty}\int_{-\infty}^{\infty} \mathrm{d}S_1\,\mathrm{d}S_2\,\mathrm{d}S_3\,p(S_1,S_2,S_3) \cdot \log\left[\frac{p(S_1,S_2,S_3)p(S_1,S_3)p(S_1)}{p(S_1,S_2)p(S_3)p(S_1,S_3)p(S_1)}\right], \tag{3.4}$$

$$I(S_1S_2;S_3) = \int_{-\infty}^{\infty}\int_{-\infty}^{\infty}\int_{-\infty}^{\infty} \mathrm{d}S_1\,\mathrm{d}S_2\,\mathrm{d}S_3\,p(S_1,S_2,S_3) \cdot \left(\log\left[\frac{p(S_1,S_3)}{p(S_3)p(S_1)}\right] + \log\left[\frac{p(S_1,S_2,S_3)p(S_1)}{p(S_1,S_2)p(S_1,S_3)}\right]\right), \tag{3.5}$$

$$I(S_1S_2;S_3) = I(S_1;S_3) + \int_{-\infty}^{\infty}\int_{-\infty}^{\infty}\int_{-\infty}^{\infty} \mathrm{d}S_1\,\mathrm{d}S_2\,\mathrm{d}S_3\,p(S_1,S_2,S_3) \cdot \log\left(\frac{p(S_2,S_3|S_1)}{p(S_2|S_1)p(S_3|S_1)}\right), \tag{3.6}$$

so that

$$I(S_1S_2;S_3) = I(S_1;S_3) + I(S_2;S_3|S_1). \tag{3.7}$$

This has been proven for arbitrary S_1, S_2, S_3 so we can also write

$$I(S_2S_1;S_3) = I(S_2;S_3) + I(S_1;S_3|S_2). \tag{3.8}$$

However an examination of (3.3) reveals that $I(S_1S_2;S_3) = I(S_2S_1;S_3)$ so that equating (3.7) and (3.8) yields

$$I(S_1;S_3) = I(S_2;S_3) + I(S_1;S_3|S_2) - I(S_2;S_3|S_1). \tag{3.9}$$

Swapping S_1 and S_3 gives

$$I(S_1;S_3) = I(S_1;S_2) + I(S_1;S_3|S_2) - I(S_1;S_2|S_3). \tag{3.10}$$

Our Markovian assumption means that

$$p(S_1,S_2,S_3) = p(S_3|S_2)p(S_2|S_1)p(S_1) \tag{3.11}$$

and

$$p(S_1,S_2) = p(S_2|S_1)p(S_1). \tag{3.12}$$

Thus

$$p(S_1,S_3|S_2) = \frac{p(S_1,S_2,S_3)}{p(S_2)} = \frac{p(S_3|S_2)p(S_2|S_1)p(S_1)}{p(S_2)} \tag{3.13}$$

or

$$p(S_1,S_3|S_2) = p(S_3|S_2)p(S_1|S_2), \tag{3.14}$$

so that S_1 and S_3 are independent when conditioned on S_2. Accordingly, $I(S_1;S_3|S_2)$ in (3.9) and (3.10) will be zero.

Remembering that conditional mutual information is always greater than or equal to zero (Gray [42, Lemma 2.5.4]) we can conclude from (3.10) and (3.9) for $I(\boldsymbol{S}_1;\boldsymbol{S}_3)$ that

$$I(\boldsymbol{S}_1;\boldsymbol{S}_3) \leq I(\boldsymbol{S}_2;\boldsymbol{S}_3), \tag{3.15}$$

and

$$I(\boldsymbol{S}_1;\boldsymbol{S}_3) \leq I(\boldsymbol{S}_1;\boldsymbol{S}_2). \tag{3.16}$$

End of Proof

This proposition is the information theoretic equivalent of the aphorism that a chain is as strong as its weakest link.

3.2 Upper Limit to Information Performance

In a wave-based positioning system the information will be transmitted over the physical communications channel. This channel will have a finite capacity measured in terms of bit/s. Denote this capacity limit as C. We can now state the following result.

Proposition 3.2 *Consider a positioning system with a capacity bound, C, on the physical communications channel. Then the performance P_T must be less than or equal to C for error free communications.*

Proof

A positioning system can be seen as a sequential set of operations on random vectors:

$$\boldsymbol{X} \rightarrow \boldsymbol{\Xi} \rightarrow \boldsymbol{\Phi} \rightarrow \boldsymbol{Y} \rightarrow \widehat{\boldsymbol{X}}. \tag{3.17}$$

Each of the transformations depend only the immediately preceding variable: it is a Markov chain. Accordingly Proposition 3.1 can be applied directly to this case:

$$I(\boldsymbol{X};\widehat{\boldsymbol{X}}) \leq I(\boldsymbol{\Xi};\widehat{\boldsymbol{X}}) \tag{3.18}$$

and

$$I(\boldsymbol{\Xi};\widehat{\boldsymbol{X}}) \leq I(\boldsymbol{\Xi};\boldsymbol{\Phi}), \tag{3.19}$$

so that

$$I(\boldsymbol{X};\widehat{\boldsymbol{X}}) \leq I(\boldsymbol{\Xi};\boldsymbol{\Phi}). \tag{3.20}$$

But $I(\boldsymbol{\Xi};\boldsymbol{\Phi})$ is the information transferred in the spatial communications channel. Then, by Shannon's Theorem [103][Theorem 11], the instantaneous information performance and the information performance will be bounded by the capacity C of the channel.

End of Proof

Care must be taken in interpreting the concept of error free communications for a positioning system. This does not mean that position can be measured with infinite precision, merely that a certain level of accuracy is achievable. It is a further issue as to whether a given level of accuracy is practically achievable (see Chap. 5 and 6).

One motivation for clearly stating the theoretical limit on the information performance of positioning systems is to stimulate interest in improving the performance of positioning systems. This can be compared to communications systems: until recently, communications systems operated well below their Shannon capacity. But over the last decade, approaches such as trellis coding [117] have allowed a significant step towards the practical achievement of the limit. This has been most spectacular in modem design for telecommunications channels resulting in dramatic increases in data rate [88]. Similarly a clear understanding of the upper limit to positioning systems should provide an impetus to improve the performance of these systems. Of course physical coding places constraints on the type of coding schemes that can be implemented, but performance enhancement techniques are still possible. There is a detailed discussion of this issue in Sect. 6.7.

When the spatial channel is the limiting factor, the capacity can be defined in terms of bandwidth and signal-to-noise ratio, according to Shannon's capacity theorem [103] or a suitable generalisation. The generalisation is necessary because the physical channel may be multi-link or multi-dimensional [54].

For a one-dimensional positioning system, operating on a channel with additive, white gaussian noise, the normal Shannon capacity limit can be applied: the capacity is given by

$$C = B \log \left(1 + \frac{S_p}{n}\right) \tag{3.21}$$

where B is the bandwidth, S_p is the signal power and n is the total noise within the channel bandwidth [114].

This equation must be generalised for a multi-dimensional channel. For example, consider a two-dimensional optical system that is being used to measure position (e.g. star tracker or spatial acquisition systems for optical links [125]). The following argument is based on the work of Cox and Sheppard [1] [18].

For a one-dimensional system, with bandwith B and duration T, the number of degrees of freedom is $2TB + 1$. The addition of the unity factor is a necessary correction for small duration bandwidth products. In the case of a two-dimensional optical system the number of degrees of freedom will be (ignoring polarisation)

$$N_f = (2L_x B_x + 1)(2L_y B_y + 1)(2TB + 1) \tag{3.22}$$

where B_x and B_y are the spatial bandwidths; L_x and L_y are the dimensions of the field of view.

[1] This derivation is only approximate. A further line of research would be to provide a rigourous derivation of the multi-dimensional channel capacity by following the arguments used for establishing single channel capacity.

If the noise is assumed to be additive and uncorrelated, then the number of distinguishable states will be given by $[\frac{1}{n}(S_p + n)]^{\frac{1}{2}}$. This means that the total number of messages will be

$$N_m = \left[\frac{(S_p + n)}{n}\right]^{\frac{N_f}{2}}. \tag{3.23}$$

As each state is equiprobable, the information content will be $\log N_m$, so the capacity will be

$$C = \frac{1}{2T}(2L_xB_x + 1)(2L_yB_y + 1)(2TB + 1)\log\left(1 + \frac{S_p}{n}\right) \tag{3.24}$$

Cox and Sheppard derived this equation to understand the effects of super resolution in optical systems. Super resolution will be discussed in more detail in Chap. 6.

Equation (3.24) can be readily applied to more conventional positioning systems such as radars and sonars (see the example below). Here again super resolution effects can be exploited. But the important factor to keep in mind is that the channel capacity cannot be exceeded.

Example 3.3

The following is an example of a capacity calculation. Consider a simple polar radar system which measures range by means of pulses and angle using a phased array antenna. Suppose that the pulse bandwidth is small compared to the carrier frequency, the directivity of the antenna is high and the field of view is relatively small. The communications channel will be two-dimensional (time and angle).

In order to make a capacity calculation we need to interpret L_x and B_x (see (3.24)) for our antenna system. In a normal communications system the receiver can be represented as in Fig. 3.1A. Here $u(t)$ is the input signal, $h(t)$ is the impulse response of the system and $w(t)$ is the output signal. If the system is linear then $w(t)$ will be the convolution of $u(t)$ and $h(t)$, or using the convolution theorem

$$W(f) = H(f)U(f) \tag{3.25}$$

where the W, H and U are the Fourier transforms of w, h and u respectively. The bandwidth is normally determined from measurements of $|H(f)|^2$, so solving for $H(f)$ gives

$$|H(f)|^2 = \frac{|W(f)|^2}{|U(f)|^2}. \tag{3.26}$$

If we feed into the input a delta function, $\delta(t)$, then this equation becomes

$$|H(f)|^2 = |W(f)|^2 \tag{3.27}$$

thus allowing a very simple determination of the frequency response.

Now consider an antenna system (see Fig. 3.1B). Here the input signal is $f(\theta)$, the impulse response of the antenna system is $m(\theta)$ and the output signal

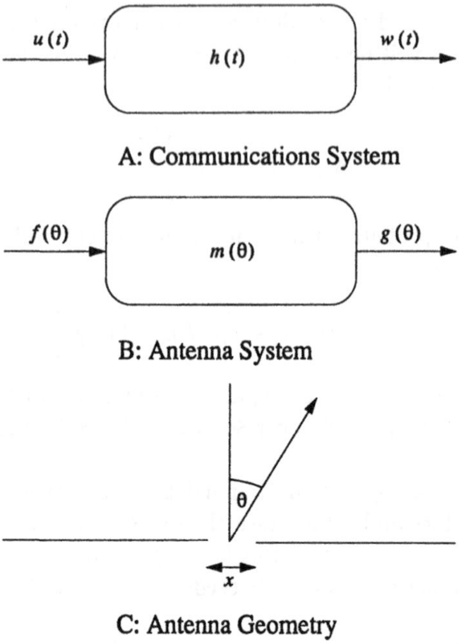

Fig. 3.1. Diagram of Communication and Antenna System

$g(\theta)$. From Bracewell [10, Chap. 13] we have that in the far field, the electric field distribution along the antenna aperture will be related to $f(\theta)$ by

$$E\left(\frac{x}{\lambda}\right) = \int_{-\infty}^{\infty} d\theta\, f(\theta) \exp\left(j2\pi \left(\frac{x}{\lambda}\right) \sin\theta\right) \tag{3.28}$$

where x is the distance along the aperture and θ is is measured from the perpendicular (see Fig. 3.1C), and λ is the wavelength.

We have assumed that the signal is a narrow band, so we can set λ equal to the carrier wavelength. In the case of high directivity and small angle of view, then $\sin\theta \simeq \theta$ and $\frac{x}{\lambda}$ and θ will form a Fourier pair [10]:

$$F\left(\frac{x}{\lambda}\right) = \int_{-\infty}^{\infty} d\theta\, f(\theta) \exp\left(j2\pi \left(\frac{x}{\lambda}\right) \theta\right) \tag{3.29}$$

and

$$f(\theta) = \int_{-\infty}^{\infty} d\left(\frac{x}{\lambda}\right) F\left(\frac{x}{\lambda}\right) \exp\left(-j2\pi \left(\frac{x}{\lambda}\right) \theta\right) \tag{3.30}$$

where as before the capitalised letter indicate the Fourier domain (e.g. $F(\frac{x}{\lambda})$ is the Fourier transform of $f(\theta)$).

In order to measure the impulse response of our antenna system we can set $f(\theta)$ equal to $\delta(\theta)$, which corresponds physically to a plane wave whose angle of incidence is zero. In this case, by analogy with (3.27) we will have that

Table 3.1. Parameters for antenna system

Parameter	Value
θ_0	30 degrees
Aperture size	20 metres
Signal to Noise	-10 dB
Bandwidth	29 KHz
Centre Frequency	300 MHz

$|M(\frac{x}{\lambda})|^2 = |G(\frac{x}{\lambda})|^2$, so that in order to calculate $|M(\frac{x}{\lambda})|^2$ we only need to work out $|G(\frac{x}{\lambda})|^2$, the response of the system to the delta function.

A plane wave will uniformly excite the aperture. Now the antenna system may further process this response to weight different parts of the aperture differently, but from the information theoretical point of view, this further processing will not change the capacity. Accordingly we have that $|G(\frac{x}{\lambda})|^2$ will be constant over the aperture and zero outside the aperture. The 'bandwidth' of this response will be

$$B_x = \frac{a}{\lambda} \tag{3.31}$$

where a is the aperture width.

The dimension of view will be the total scan angle. Suppose the system scans symmetrically on either side of $\theta = 0$. Denote the maximum angular extent of the scan as $\pm\theta_0$, so that the total scan angle is $2\theta_0$. Provided $\sin\theta_0 \simeq \theta$ our requirement for small θ will be satisfied. Accordingly

$$L_x = 2\theta_0, \tag{3.32}$$

so that

$$C = \frac{1}{2T}\left(\frac{4\theta_0 a}{\lambda} + 1\right)(2BT + 1)\log\left(1 + \frac{S_p}{n}\right). \tag{3.33}$$

Suppose the system parameters are as shown in Table 3.1

If these parameters are inserted into (3.24) the result is a capacity of 171 kbit/s. By calculating the optimal accuracy (i.e. the accuracy that optimises the rate of useful information gain) it is possible to convert this capacity into an estimate of the maximum measurement rate at the optimal accuracy. Note that if there was an omnidirectional antenna the capacity would only be 4 kbit/s, so in this case the antenna can provide the opportunity for a twenty-fold increase in performance. This increase in performance should be realizable as a much higher measurement rate for the system.

In a practical radar system, there will be many sources transmitting (i.e. objects reflecting the radar signal), so the analysis of capacity would have to be extended to include multiple sources. Many commercial positioning systems are either self positioning or use some multiplexing scheme so that the single source analysis is sufficient. □

The above is an example of a multi-dimensional system. Now consider an example of a multi-link system.

Example 3.4 Suppose there is a radial-radial system that eliminates multiple solutions by confining the region of interest (see Fig. 3.2). Suppose the radius is measured in each case by a pulsed radar system operating in a bandwidth of 4 MHz. In this case the communications channel is multi-link, with each radar system seen as a separate independent communications link. If signal-to-noise ratio considerations dictate the links have a capacity of 1 kbit/s then the performance limit in the world frame will be 2 kbit/s. □

3.3 Invariance

In Sect. 3.2 we established an upper bound on performance such that $I(\boldsymbol{X};\widehat{\boldsymbol{X}}) \leq I(\boldsymbol{\Xi};\boldsymbol{\Phi})$. In certain circumstances it is possible to show that $I(\boldsymbol{X};\widehat{\boldsymbol{X}}) = I(\boldsymbol{\Xi};\boldsymbol{\Phi})$. This property is called invariance. We will show that this will be achieved if $I(\boldsymbol{X};\boldsymbol{Y}) = I(\boldsymbol{\Xi};\boldsymbol{\Phi})$ and $I(\boldsymbol{X};\boldsymbol{Y}) = I(\boldsymbol{X};\widehat{\boldsymbol{X}})$. The first condition is called *device invariance* while the second is called *estimation invariance*.

3.3.1 Device Invariance

If the average mutual information in the world and communications frames are equal, the positioning system will be called device invariant. Invariant systems are much easier to analyse. In addition the performance of such systems is invariant and so a more meaningful parameter.

An important circumstance when systems are device invariant is outlined below:

Proposition 3.5

Consider mappings $\boldsymbol{\xi} = \boldsymbol{g}(\boldsymbol{x})$ *and* $\boldsymbol{\phi} = \boldsymbol{f}^{-1}(\boldsymbol{y})$. *Assume that* $\boldsymbol{g}$ *and* $\boldsymbol{f}^{-1}$ *are 1:1 mappings, with* $\boldsymbol{g}$ *and* $\boldsymbol{f}^{-1}$ *independent in the sense that*

$$\frac{\partial g_i}{\partial y_{jk}} = 0 \; ; \; i = 1, \ldots, n; \; j = 1, \ldots, n, \; k = 1, \ldots, K \tag{3.34}$$

and

$$\frac{\partial f_i^{-1}}{\partial x_{jk}} = 0 \; ; \; i = 1, \ldots, n; \; j = 1, \ldots, n, \; k = 1, \ldots, K \tag{3.35}$$

then $I(\boldsymbol{X};\boldsymbol{Y}) = I(\boldsymbol{\Xi};\boldsymbol{\Phi})$.

A positioning system whose mappings satisfy the assumptions of this proposition will be referred to as *well-defined.* A proof of this proposition may be found in Appendix C.

The requirements that the cross derivatives are zero (see (3.34) and (3.35)) will normally be fulfilled for a real-world positioning system. This is because the mapping $\boldsymbol{g} : \mathcal{W} \longrightarrow \mathcal{C}$ is a function of the geometry and so will not depend on the output of the positioning system. If the mapping $\boldsymbol{f}^{-1} : \mathcal{C} \longrightarrow \mathcal{W}$ depends on the actual position, then the designer would be better off using the known position, rather than worrying about the positioning system.

However the assumption that $\boldsymbol{f}$ and $\boldsymbol{g}$ are 1:1 mappings can be violated. There are many systems (e.g. Loran) which can yield multiple solutions. Call this type of ambiguity ***physical ambiguity*** (other types of ambiguity will be discussed in Chap. 6). When ambiguous solutions are possible the designer seeks to resolve the ambiguity. This is normally done in either of two ways:

- By constraining the region of interest so that ambiguities do not occur. In a radial-radial system (see Fig. 3.2), this could be done by constraining the targets' motion to within the shaded area, where the circular loci cross only at a single point. Proposition 3.5 can be applied directly to such a system.
- By using a communications frame with a greater dimensionality than the world frame. For example the multiple solutions in the radial-radial co-ordinate system can be eliminated by using a third measurement point. This yields a radial-radial-radial positioning system (see Fig. 3.3).

Given that the second technique is used in many tracking systems, it is necessary to generalise Proposition 3.5 to account for this case. Formally, the way that the ambiguity is resolved is by augmenting the measurement with the addition of an extra variable η in the communications frame. Normally this η will be of similar type to one of the $\xi_1, \ldots, \xi_n$ but will be independent of them. Accordingly $\boldsymbol{\xi}$ becomes $\boldsymbol{\xi}'$, where $\boldsymbol{\xi}'$ is given by

$$\boldsymbol{\xi}' = (\xi_1, \ldots, \xi_n, \eta)^{\mathrm{T}}. \tag{3.36}$$

The augmentation can be extended to $\boldsymbol{\Xi}$ and $\boldsymbol{\Phi}$. This defines an augmented communications frame. A frame augmented in this fashion is referred to as ***independently-augmented***.

The image of V_n in this $(n+1)$-dimensional augmented frame is a sub-space of dimensionality n (except perhaps for singularities). Accordingly, to provide a meaningful transformation of the p.d.f. it is necessary to augment the world frame with a dummy variable e. Let the augmented variable $\boldsymbol{x}'$ be given by

$$\boldsymbol{x}' = (x_1, \ldots, x_n, e)^{\mathrm{T}}. \tag{3.37}$$

Define the augmented p.d.f. of $\boldsymbol{X}$ as follows:

$$p'_X(\boldsymbol{X}') = p_X(\boldsymbol{X})\delta(e_1)\ldots\delta(e_K) \tag{3.38}$$

where $\delta(\cdot)$ is the Dirac delta function and e_i is the dummy variable corresponding to the ith measurement. In a similar fashion we can define the augmented quantities $\boldsymbol{Y}'$, $p'(\boldsymbol{X}', \boldsymbol{Y}')$ and $p'_Y(\boldsymbol{Y}')$, where the augmented variable corresponding to $\boldsymbol{y}$ is f.

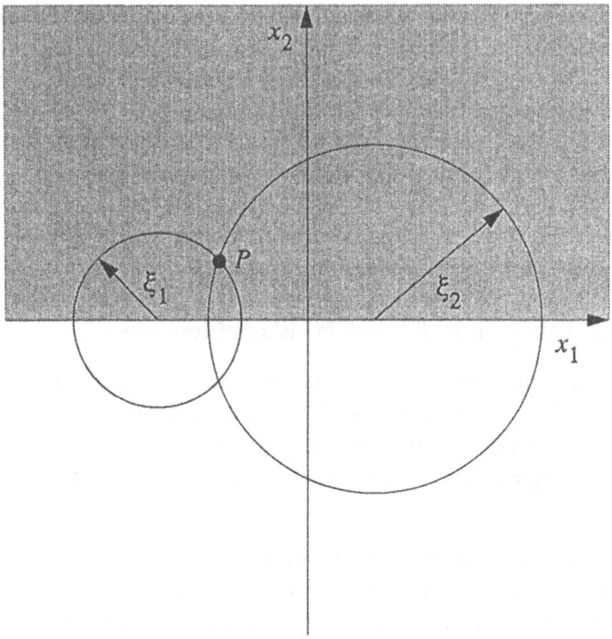

Fig. 3.2. Diagram of Constrained Radial-Radial System

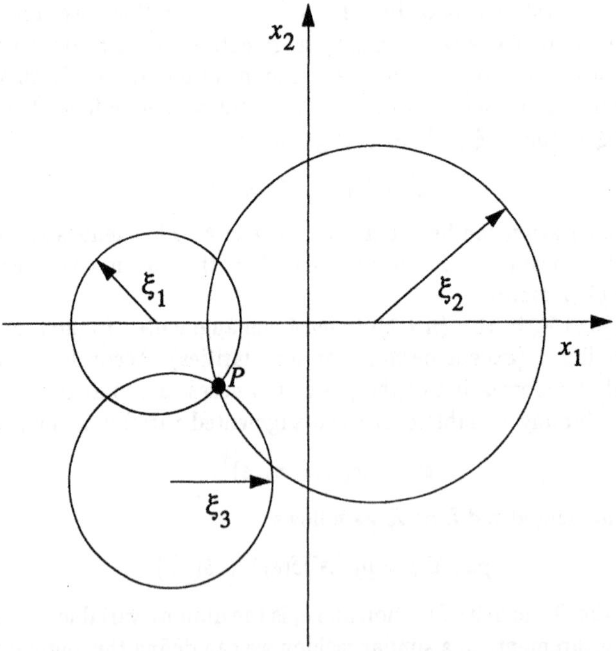

Fig. 3.3. Diagram of Radial-Radial-Radial System

These augmented variables and distributions define a mapping from the world frame to an augmented world frame. Refer to a system augmented in this fashion as *δ-augmented*.

Given these definitions, the following proposition can be stated.

Proposition 3.6 *Consider a positioning system which independently-augments the communication frame. If the mappings between the δ-augmented world frame and the independently-augmented communications frame are well-defined (in the sense of Proposition 1), then*

$$I(\boldsymbol{X};\boldsymbol{Y}) = I(\boldsymbol{\Xi}';\boldsymbol{\Phi}') \tag{3.39}$$

A proof of this proposition may be found in Appendix D. The last two propositions can also be understood by thinking of the positioning systems as a data processing system. If the transformation from $\boldsymbol{x} \longrightarrow \boldsymbol{\xi}$ and $\boldsymbol{\phi} \longrightarrow \boldsymbol{y}$ are 1:1, then no information is lost, so it is expected that the system will be invariant.

The significance of Propositions 3.5 and 3.6 is that for almost all practical positioning systems, the average mutual information is the same in the communications frame and the world frame. This allows important operations such as optimisation and capacity calculations to be carried out in the communications frame. This also considerably simplifies the mathematical analysis, and also makes the results easier to understand.

Example 3.7

Suppose a one-dimensional positioning system takes only one measurement. Assume that x is a zero-mean gaussian random variable (g.r.v.) with variance σ^2. The positioning device is $\xi = g(x) = ax; \phi = q(\xi) = \xi + \zeta;\ \phi = f(y) = ay$; where a is an arbitrary constant and ζ is a zero-mean random variable with variance τ^2.

In this case ξ will be a zero-mean g.r.v. with variance $a^2\sigma^2$. From Example 2.1, we see that the average mutual information in the communications frame is given by

$$I(\Xi;\Phi) = \frac{1}{2}\log\left(1+\frac{a^2\sigma^2}{\tau^2}\right). \tag{3.40}$$

By transforming the p.d.fs in the communications frame to the world frame it is seen readily that in the world frame y can be thought of as the sum of two zero-mean random variables x and z. The x has variance σ^2, while z has variance $\frac{\tau^2}{a^2}$. This is different from the variances of ξ and ζ, but the average mutual information in the world frame is given by

$$I(\mathbf{X};\mathbf{Y}) = \frac{1}{2}\log\left(1+\frac{a^2\sigma^2}{\tau^2}\right) \tag{3.41}$$

which is the same as in the communications frame. □

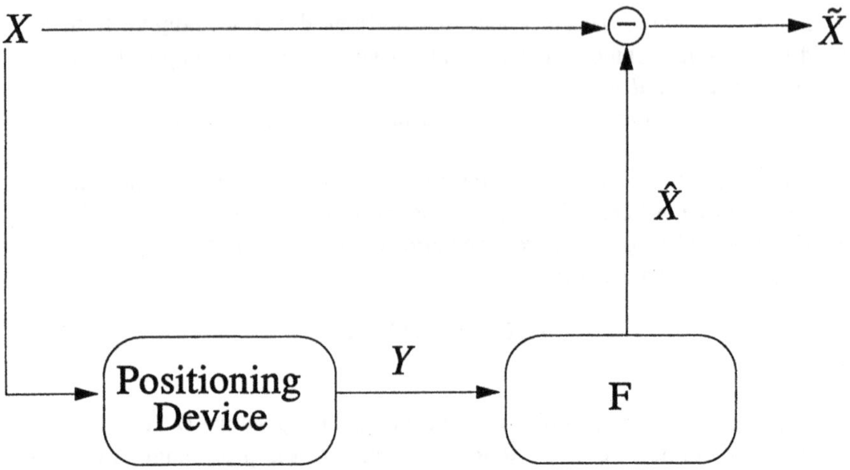

Fig. 3.4. Diagram of Estimation Process

3.3.2 Estimator Invariance

Consider the abstract representation of a positioning system shown in Fig. 3.4. Here the waveform coding/decoding and the physical coding/decoding occur in the box marked **Positioning Device**. The position vector $\boldsymbol{X}$ is estimated by the positioning device which produces a measurement vector $\boldsymbol{Y}$. This measurement vector is fed into a filter, marked **F**, which estimates the position. The output from the filter is $\widehat{\boldsymbol{X}}$. Now the normal method of designing the filter is to minimise the expectation of some cost which is a function of the difference between $\boldsymbol{X}$ and $\widehat{\boldsymbol{X}}$. Define $\widetilde{\boldsymbol{X}}$ as

$$\widetilde{\boldsymbol{X}} = \boldsymbol{X} - \widehat{\boldsymbol{X}}. \tag{3.42}$$

Typical cost functions [100] are the squared error function $\widetilde{\boldsymbol{X}}^{\mathrm{T}}\widetilde{\boldsymbol{X}}$, the absolute value of error $\sqrt{\widetilde{\boldsymbol{X}}^{\mathrm{T}}\widetilde{\boldsymbol{X}}}$ and the uniform cost function which is 0 if $\sqrt{\widetilde{\boldsymbol{X}}^{\mathrm{T}}\widetilde{\boldsymbol{X}}} < \frac{1}{2}\epsilon$ and equal to ϵ^{-1} if $\sqrt{\widetilde{\boldsymbol{X}}^{\mathrm{T}}\widetilde{\boldsymbol{X}}} \geq \frac{1}{2}\epsilon$, for some ϵ. Note that in the limit as ϵ approaches zero, the estimator that uses the uniform cost function becomes the maximum *a priori* estimator (see Sect. 6.1). For many systems these three estimators will give identical results [100].

The cost function $C(\widetilde{\boldsymbol{X}})$ to be used in the following development is [123]:

$$C(\widetilde{\boldsymbol{X}}) = -\log(p_{\widetilde{X}}(\widetilde{\boldsymbol{X}})) \tag{3.43}$$

so that the expected value of this cost function is the entropy of $\widetilde{X}$. Call an estimator that uses this cost function the *entropy-error estimator.* With reference to Fig. 3.4, the aim is to choose F to minimise the entropy of $\widetilde{X}$. For gaussian systems this cost function will give the same result as the more conventional cost functions (see Sect. 7.3).

Now let us define a bound on the entropy of $\widetilde{X}$. Following the method of Weidermann and Stear [123], we can prove the following proposition.

Proposition 3.8 *The entropy of $\widetilde{X}$ has a lower bound of $H(X) - I(X;Y)$. If an estimator, F, which achieves this lower bound exists, then this estimator will be called the optimum estimator and be denoted $\widehat{F}$. The optimum estimator will cause the error vector $\widetilde{X}$ to be independent of the measurement vector Y. In addition, for the optimum estimator $I(\widetilde{X};Y) = 0$.*

Proof

Using (3.42) and the definition of F we have that

$$X = \widetilde{X} + \widehat{X} \tag{3.44}$$

and

$$p_{\widetilde{X}Y}(\widetilde{X}, Y) = p_{XY}(\widetilde{X} + F(Y), Y) \tag{3.45}$$

so that

$$H(\widetilde{X}, Y) = H(X, Y). \tag{3.46}$$

Now consider the average mutual information between the error vector and Y:

$$I(\widetilde{X};Y) = H(\widetilde{X}) + H(Y) - H(\widetilde{X}, Y) + (H(X) - H(X)), \tag{3.47}$$

so using (3.46)

$$I(\widetilde{X};Y) = H(\widetilde{X}) - H(X) + H(X) + H(Y) - H(X, Y), \tag{3.48}$$

or

$$I(\widetilde{X};Y) = H(\widetilde{X}) - H(X) + I(X;Y), \tag{3.49}$$

so that

$$H(\widetilde{X}) = H(X) - I(X;Y) + I(\widetilde{X};Y). \tag{3.50}$$

The average mutual information is always greater than or equal to zero so that $I(\widetilde{X};Y) \geq 0$ and

$$H(\widetilde{X}) \geq H(X) - I(X;Y). \tag{3.51}$$

This provides a lower bound on the entropy of the error vector.

Consider (3.50). On the right hand side only $I(\widetilde{X};Y)$ depends on $\widetilde{X}$ so that the only way to achieve the lower bound is if $I(\widetilde{X};Y) = 0$.

Accordingly, if the optimal estimator exists we have that $I(\widetilde{X};Y) = 0$. Given that the average mutual information between two vectors is zero if and only if the vectors are independent we can also state that $\widetilde{\boldsymbol{X}}$ and $\boldsymbol{Y}$ are independent. This result makes sense. If there is residual information between the error and the data, then the data should be processed further. Note that this requirement is more stringent than estimators which only require that the error and the data are orthogonal, which indicates that in certain cases the entropy-error estimator may give superior performance, though it may be less tractable mathematically.

End of Proof

Using the above result we can prove the following important proposition.

Proposition 3.9 *For a positioning system where the optimal estimator $\widehat{\boldsymbol{F}}$ exists, we have that $I(\boldsymbol{X};\boldsymbol{Y}) = I(\widehat{\boldsymbol{X}};\boldsymbol{X})$.*

Proof

For the optimal estimator we have (see Proposition 3.8),

$$I(\boldsymbol{X};\boldsymbol{Y}) = H(\boldsymbol{X}) - H(\widetilde{\boldsymbol{X}}) \tag{3.52}$$

and

$$I(\widetilde{\boldsymbol{X}};\boldsymbol{Y}) = 0. \tag{3.53}$$

From the generalised data processing theorem we have the result that, for an arbitrary vector $\boldsymbol{T}$ and an arbitrary deterministic function $\boldsymbol{D}$, we have that $I(\boldsymbol{T};\boldsymbol{Y}) \geq I(\boldsymbol{T};\boldsymbol{D}(\boldsymbol{Y}))$. Accordingly

$$I(\widetilde{\boldsymbol{X}};\widehat{\boldsymbol{X}}) = I(\widetilde{\boldsymbol{X}};\boldsymbol{F}(\boldsymbol{Y})) \leq I(\widetilde{\boldsymbol{X}};\boldsymbol{Y}) = 0. \tag{3.54}$$

But average mutual information is always greater than zero so that

$$I(\widetilde{\boldsymbol{X}};\widehat{\boldsymbol{X}}) = 0. \tag{3.55}$$

This means that for the optimal estimator, $\boldsymbol{X}$ is the sum of two independent random vectors $\widetilde{\boldsymbol{X}}$ and $\widehat{\boldsymbol{X}}$, so that

$$I(\boldsymbol{X};\widehat{\boldsymbol{X}}) = H(\boldsymbol{X}) - H(\widetilde{\boldsymbol{X}}). \tag{3.56}$$

By reference to (3.52) we can write

$$I(\boldsymbol{X};\widehat{\boldsymbol{X}}) = I(\boldsymbol{X};\boldsymbol{Y}). \tag{3.57}$$

End of Proof

The importance of this proposition, together with Propositions 3.5 and 3.6, is that a system which uses the optimum estimator will be invariant. This means that maximising the information flow in the communications frame will

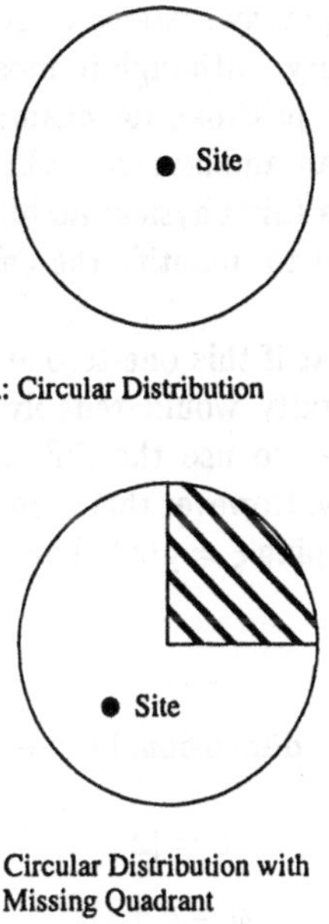

Fig. 3.5. Example of non-optimal estimator

automatically optimise the overall system performance. This is not necessarily so for a system that does not use an optimum estimator. The property of invariance will be assumed when discussing system optimisation in Chap. 4

Example 3.10

Consider a polar positioning system with a uniform circular *a priori* p.d.f. in the communications frame (see Fig. 3.5A). If the estimator is optimal the best place to put the positioning device is at the centre of the p.d.f. (see Example 4.1). However, consider an estimator which ignores any position measurement in one quadrant. The best place to put the positioning device will no longer be the centre of the p.d.f., but somewhere in the other three quadrants (see Fig. 3.5B). □

3.4 Physical Ambiguity

In Sect. 3.3 it was shown that if there is a one-to-one mapping between the world frame and the communications frame, then the average mutual information in the world frame would equal the average mutual information in the communications frame and so the system would be considered to be invariant.

It was also pointed out that in many systems there is not a one-to-one mapping, which causes physical ambiguity. Although in most systems this ambiguity can be overcome, it is not always possible, for example a GPS receiver operating in an urban area can return an ambiguous reading if the line of sight to some of the satellites is blocked. As this physical ambiguity will degrade the performance of a system it is useful to quantify the effect of physical ambiguity on the information performance.

An initial thought was that if this one-to-one mapping implies that $I(\boldsymbol{x}, \boldsymbol{y}) = I(\boldsymbol{\xi}, \boldsymbol{\phi})$ then ambiguity would reliably evidence itself as departure from this equivalence: that is, to use the difference $I(\boldsymbol{x}, \boldsymbol{y}) - I(\boldsymbol{\xi}, \boldsymbol{\phi})$ as an indicator of physical ambiguity. However this is not a good measure of ambiguity because y depends on the mapping $\xi = g(x)$. This can be seen from the following example.

Example 3.11

Consider the following one-dimensional system, which has a symmetric positional p.d.f. $p_x(x)$ and

$$\xi = |x|, \tag{3.58}$$

$$\phi = \xi + \zeta, \tag{3.59}$$

$$y = \phi, \tag{3.60}$$

where ζ is a zero-mean gaussian distributed noise with variance σ.

Now given that the positional p.d.f. is symmetric about zero, one would expect that the two-fold ambiguity would lead to one bit loss in information.

But, consider $I(x, y) - I(\xi, \phi)$ for this case. We have that

$$\begin{aligned} I(x,y) - I(\xi,\phi) &= \int_{-\infty}^{\infty}\int_{-\infty}^{\infty} \mathrm{d}x\ \mathrm{d}y\, p_x(x)\frac{1}{\sqrt{2\pi}}e^{-\frac{(y-|x|)^2}{2\sigma^2}} \log\left(\frac{1}{\sqrt{2\pi}}e^{-\frac{(y-|x|)^2}{2\sigma^2}}\right) - \\ &\quad \int_{-\infty}^{\infty}\int_{0}^{\infty} \mathrm{d}\xi\ \mathrm{d}\phi\, 2p_x(\xi)\frac{1}{\sqrt{2\pi}}e^{-\frac{(\phi-\xi)^2}{2\sigma^2}} \log\left(\frac{1}{\sqrt{2\pi}}e^{-\frac{(\phi-\xi)^2}{2\sigma^2}}\right) \end{aligned} \tag{3.61}$$

so, noting that $p_x(x)$ is symmetric about $x = 0$, the second integral will equal the first integral and we have

$$I(x,y) - I(\xi,\phi) = 0 \tag{3.62}$$

and the system is invariant. □

This example shows that we must search for a better measure of the degree of ambiguity.

A definition which was inspired by Shaw's definition of information flow for chaotic maps [104] is as follows:

Define the physical ambiguity measure as

$$P_a = H(\boldsymbol{u}) - H(\boldsymbol{x}), \tag{3.63}$$

where

$$\boldsymbol{u} = \boldsymbol{f}(\boldsymbol{g}(\boldsymbol{x})). \tag{3.64}$$

This measures, in units of bits, the degree of ambiguity involved with the physical coding/decoding of a system, weighted by the positional p.d.f.

The most common situation will be that $\boldsymbol{g}$ is a one to one mapping and f is the inverse. In this case, $\boldsymbol{u} = \boldsymbol{x}$ and $P_a = 0$. However, more generally, P_a will depend upon the positional p.d.f. This can be seen from the following example.

Example 3.12 Consider the system described in Example 3.11, except that $p_x(x)$ is given by

$$p_x(x) = \begin{cases} 1 & \text{if } \alpha - 1 < x < \alpha \\ 0 & \text{elsewhere} \end{cases} \tag{3.65}$$

where α is a parameter that must lie between zero and one half.

In this case we have that $u = |x|$, and that

$$p(u) = \begin{cases} 0 & \text{if } x < 0 \\ 2 & \text{if } 0 < x < \alpha \\ 1 & \text{if } \alpha < x < 1 - \alpha. \end{cases} \tag{3.66}$$

So in this case we have that

$$P_a = -2\alpha \log_2 2 \; ; \; 0 < \alpha < \frac{1}{2}. \tag{3.67}$$

If $\alpha = 0$, the positional p.d.f. lies entirely on the negative x-axis, and the ambiguity will have no effect, so it is to be expected that $P_a = 0$. On the other hand, if $\alpha = \frac{1}{2}$ then the positional p.d.f. will be symmetric about $x = 0$ and there should be a one bit loss of information ($P_a = -1$). □

4. System Optimisation

This chapter considers methods of optimising positioning systems. First the technique for optimising the configuration of a device is presented and a method of comparing several systems is discussed. Then a method of deciding on the optimal measurement strategy is evolved. Next a general result, the independent error approximation, is derived. This allows a reduction in the dimensionality of the optimisation problems. Finally a method of defining and optimising classes of systems is developed.

4.1 Optimal Configuration

The performance of a positioning device will normally depend on the source statistics, the measurement vector and a number of parameters that are under the control of the system designer. Let these control parameters be $u_1, \ldots, u_L$, where L is the number of parameters.

Accordingly the optimisation problem is to maximise $P_T(u_1, \ldots, u_L)$, perhaps subject to constraints on the u_i. In this case P_T is defined as

$$P_T(u_1, \ldots, u_L) = \frac{1}{T} I(\boldsymbol{X}; \boldsymbol{Y}) \tag{4.1}$$

where T is the duration of the observation. $\boldsymbol{Y}$ will normally be a function of the u_i. $\boldsymbol{X}$ is not necessarily a function of the u_i, but may be, as the measurement strategy may indirectly depend on the control parameters. Note that in this chapter, estimator invariance is assumed so that $I(\boldsymbol{X}; \widehat{\boldsymbol{X}}) = I(\boldsymbol{X}; \boldsymbol{Y})$.

A common sort of control parameter arises when the Jacobian of the transformation depends on several co-ordinate points. For example the Jacobian for a polar radar system will be dependent on where the radar is placed relative to the positional p.d.fs. Similarly the performance of a radial-radial system will depend on where the two ranging devices are placed. This problem is critical to the design of vehicle tracking systems and other wide area positioning systems which may require the placement of tens of reference sites.

In general, this sort of problem would have to be solved using numerical techniques. However by making a number of assumptions it is possible to gain an analytic insight into the optimal configuration.

Assume that each measurement is independent with the same statistics and that in the communications frame the measurement error is additive and much smaller than the *a priori* uncertainty in position. The transformation between world and communications frames is assumed to be invariant.

Suppose that the positioning system depends on L coordinate points: $\boldsymbol{u}_1, \ldots, \boldsymbol{u}_L$. These are measured in the world frame. Then we have that the average mutual information is given by $I(\boldsymbol{\Xi}; \boldsymbol{\Phi})$ where I has a dependence on $\boldsymbol{u}_1, \ldots, \boldsymbol{u}_L$. The optimisation problem is to determine the values of $\boldsymbol{u}_1, \ldots, \boldsymbol{u}_L$ that maximise I. Given the assumption that each measurement is independent we may replace the measurement vectors $\boldsymbol{\Xi}$ and $\boldsymbol{\Phi}$ by single measurements $\boldsymbol{\xi}$ and $\boldsymbol{\phi}$.

Given that the transformation is invariant we can write

$$\frac{I(\boldsymbol{x}; \boldsymbol{y})}{T} = \frac{I(\boldsymbol{\xi}; \boldsymbol{\phi})}{T}. \tag{4.2}$$

The measurement error being additive and independent allows us to separate the average mutual information into the difference between entropy of $\boldsymbol{\phi}$ and the entropy of the measurement error (Jones [53, page 149]):

$$I(\boldsymbol{x}; \boldsymbol{y}) = \frac{H(\boldsymbol{\phi}; \boldsymbol{u}_1, \ldots, \boldsymbol{u}_L) - H(\boldsymbol{\zeta})}{T} \tag{4.3}$$

where $\boldsymbol{\zeta}$ is the measurement error, and the concatenation of "$H(\boldsymbol{\phi})$" and "$; \boldsymbol{u}_1, \ldots, \boldsymbol{u}_L$" indicates the explicit dependence on $\boldsymbol{u}_1, \ldots, \boldsymbol{u}_L$

It is assumed that $H(\boldsymbol{\zeta})$ will not be a function of $\boldsymbol{u}_1, \ldots, \boldsymbol{u}_L$. In this case the optimal value of I can be determined by considering only $H(\boldsymbol{\phi}; \boldsymbol{u}_1, \ldots, \boldsymbol{u}_L)$. If the measurement variance is much less than the positional p.d.f. variance we can write

$$H(\boldsymbol{\phi}; \boldsymbol{u}_1, \ldots, \boldsymbol{u}_L) \simeq H(\boldsymbol{\xi}; \boldsymbol{u}_1, \ldots, \boldsymbol{u}_L). \tag{4.4}$$

Then

$$H(\boldsymbol{\phi}; \boldsymbol{u}_1, \ldots, \boldsymbol{u}_L) \simeq - \int_{U_n} \mathrm{d}\boldsymbol{\xi}\, p_\xi(\boldsymbol{\xi}; \boldsymbol{u}_1, \ldots, \boldsymbol{u}_L) \log(p_\xi(\boldsymbol{\xi}; \boldsymbol{u}_1, \ldots, \boldsymbol{u}_L)), \tag{4.5}$$

so explicitly writing the dependence on coordinate transformation gives

$$H(\boldsymbol{\phi}; \boldsymbol{u}_1, \ldots, \boldsymbol{u}_L) \simeq - \int_{V_n} \mathrm{d}\boldsymbol{x}\, p_x(\boldsymbol{x}) \log\left(\frac{p_x(\boldsymbol{x})}{J(\boldsymbol{x}; \boldsymbol{u}_1, \ldots, \boldsymbol{u}_L)}\right), \tag{4.6}$$

where $J(\boldsymbol{x}; \boldsymbol{u}_1, \ldots, \boldsymbol{u}_L)$ denotes the absolute value of the Jacobian

$$J(\boldsymbol{x}; \boldsymbol{u}_1, \ldots, \boldsymbol{u}_L) = |\det[\boldsymbol{J}_m(\boldsymbol{x}; \boldsymbol{u}_1, \ldots, \boldsymbol{u}_L)]| \tag{4.7}$$

where $\boldsymbol{J}_m(\boldsymbol{x}; \boldsymbol{u}_1, \ldots, \boldsymbol{u}_L)$ is the Jacobian matrix defined by

$$[\boldsymbol{J}_m(\boldsymbol{x}; \boldsymbol{u}_1, \ldots, \boldsymbol{u}_L)]_{ij} = \frac{\partial g_i(\boldsymbol{x}; \boldsymbol{u}_1, \ldots, \boldsymbol{u}_L)}{\partial x_j} \tag{4.8}$$

where $i = 1, \ldots, n$ and $j = 1, \ldots, n$. So

$$H(\boldsymbol{\phi}; \boldsymbol{u}_1, \ldots, \boldsymbol{u}_L) \simeq -\int_{V_n} \mathrm{d}\boldsymbol{x}\, p_{\boldsymbol{x}}(\boldsymbol{x}) \log(p_{\boldsymbol{x}}(\boldsymbol{x})) + \int_{V_n} \mathrm{d}\boldsymbol{x}\, p_{\boldsymbol{x}}(\boldsymbol{x}) \log(J(\boldsymbol{x}; \boldsymbol{u}_1, \ldots, \boldsymbol{u}_L)). \tag{4.9}$$

The first integral on the right hand side of the above equation is independent of $\boldsymbol{u}_1, \ldots, \boldsymbol{u}_L$, so the problem of optimising $I(\boldsymbol{\xi}; \boldsymbol{\phi})$ reduces to finding the maximum of

$$M_{\boldsymbol{x}}(\boldsymbol{u}_1, \ldots, \boldsymbol{u}_L) = \int_{V_n} \mathrm{d}\boldsymbol{x}\, p_{\boldsymbol{x}}(\boldsymbol{x}) \log(J(\boldsymbol{x}; \boldsymbol{u}_1, \ldots, \boldsymbol{u}_L)). \tag{4.10}$$

From this, it can be seen that the values of $\boldsymbol{u}_1, \ldots, \boldsymbol{u}_L$ are adjusted to maximise $M_{\boldsymbol{x}}$, which represents the expected value of the log of the Jacobian.

Example 4.1

Consider a polar positioning system. Assume that the *a priori* p.d.f. in the world frame has radial symmetry around the origin, the volume of interest is all of space, each measurement is independent and the measurement error is additive and much smaller than the *a priori* uncertainty.

Suppose the system designer has to decide the optimal location for the positioning system. Denote (u_1, u_2) as the origin of the positioning system in the world frame. According to the analysis above, the optimisation problem is equivalent to finding the maximum of the following integral:

$$M_{\boldsymbol{x}}(u_1, u_2) = \int\int_{V_n} \mathrm{d}x_1\, \mathrm{d}x_2\, p_{\boldsymbol{x}}(x_1, x_2) \log(J(x_1 - u_1, x_2 - u_2)). \tag{4.11}$$

Assuming no local maxima, the maximum value of this integral will be given by the solution to the following two equations:

$$\frac{\partial M_{\boldsymbol{x}}}{\partial u_1} = 0 \tag{4.12}$$

and

$$\frac{\partial M_{\boldsymbol{x}}}{\partial u_2} = 0. \tag{4.13}$$

For a polar co-ordinate system the mapping from world frame to the communications frame will be given by

$$\xi_1 = \sqrt{x_2^2 + x_1^2} \tag{4.14}$$

and

$$\xi_2 = \tan_2^{-1}(x_2, x_1). \tag{4.15}$$

In this case the Jacobian will be given by

$$J(x_1 - u_1, x_2 - u_2) = \frac{1}{\sqrt{(x_1 - u_1)^2 + (x_2 - u_2)^2}}. \tag{4.16}$$

Accordingly we have that

$$M_x(u_1,u_2) = \int\int_{V_n} \mathrm{d}x_1\,\mathrm{d}x_2\, p_x(x_1,x_2)\log\left[\frac{1}{\sqrt{(x_1-u_1)^2+(x_2-u_2)^2}}\right]. \quad (4.17)$$

Given that the *a priori* p.d.f. has radial symmetry, it is appropriate to use the following transformations on (4.17):

$$\rho = \sqrt{x_2^2 + x_1^2} \quad (4.18)$$

and

$$\theta = \tan_2(x_2, x_1). \quad (4.19)$$

This gives

$$M_x(u_1,u_2) = -\int_{-\pi}^{\pi}\int_0^{\infty} \mathrm{d}\theta\,\mathrm{d}\rho\, p_x(\rho)\rho\log\rho_s(\rho,\theta;u_1,u_2) \quad (4.20)$$

where ρ_s is defined by

$$\rho_s(\rho,\theta;u_1,u_2) = \sqrt{(\rho\cos\theta - u_1)^2 + (\rho\sin\theta - u_2)^2}. \quad (4.21)$$

Now consider

$$\int_{-\pi}^{\pi} \mathrm{d}\theta\, \log\rho_s(\rho,\theta;u_1,u_2). \quad (4.22)$$

This is a standard integral (see Grashteyn [41, page 528]) and can be written as

$$\int_{-\pi}^{\pi} \mathrm{d}\theta\, \log\rho_s(\rho,\theta;u_1,u_2) = \begin{cases} \pi\log\rho^2 & \text{if } \rho_u \le \rho \\ \pi\log\rho_u^2 & \text{if } \rho_u \ge \rho \end{cases} \quad (4.23)$$

where $\rho_u = \sqrt{u_1^2 + u_2^2}$.

Substituting this equation into (4.20) gives

$$M_x(\rho_u) = -\pi\int_0^{\rho_u} p_x(\rho)\rho\log\rho_u^2\,\mathrm{d}\rho - \pi\int_{\rho_u}^{\infty} p_x(\rho)\rho\log\rho^2\,\mathrm{d}\rho. \quad (4.24)$$

Because of the radial symmetry the optimisation problem reduces to solving

$$\frac{\mathrm{d}M_u(\rho_u)}{\mathrm{d}\rho_u} = 0, \quad (4.25)$$

so substituting equation (4.24) into equation (4.25) yields

$$\frac{1}{\rho_u}\int_0^{\rho_u} \mathrm{d}\rho\, p_x(\rho)\rho = 0. \quad (4.26)$$

If $p_x(\rho)$ is well behaved, the solution will be $\rho_u = 0$. Accordingly for most practical radial symmetric p.d.fs the optimal location for a polar positioning system is the origin, an intuitively satisfying conclusion. It is interesting to note that if

$$p_x(\rho) = 0 \text{ for } \rho \le \rho_c \quad (4.27)$$

where ρ_c is some finite critical radius, the optimum is achieved as long as

$$\rho_u \leq \rho_c. \tag{4.28}$$

□

A possible concern with using the maximisation of average mutual information as an optimality criterion is that $I(\boldsymbol{x}, \hat{\boldsymbol{x}})$ is insensitive to invertible transformations e.g. $I(\boldsymbol{x}, \hat{\boldsymbol{x}}) = I(\boldsymbol{x}, -\hat{\boldsymbol{x}})$. This is a valid objection if we were endeavouring to find an optimal value for $\hat{\boldsymbol{x}}$, as occurs during estimation. Indeed this is why the entropy-error estimator is used in Sect. 3.3. However in this chapter, we are using the average mutual information to evaluate different options, each of which is a viable system, rather than using it to design an estimating function.

4.2 Comparison of Different Systems

In the early stage of system design, it is often necessary to compare the performance of at least two different systems (for example, Horing[47]). This can be difficult for positioning systems, particularly if they have different measurement rates and accuracies. In general terms, the problem is to compare two systems which operate on the same source with the same measurement constraints. For System A we have

$$P_T^A = \frac{I(\boldsymbol{X}^A; \boldsymbol{Y}^A)}{T} \tag{4.29}$$

and for System B

$$P_T^B = \frac{I(\boldsymbol{X}^B; \boldsymbol{Y}^B)}{T}. \tag{4.30}$$

System A will be said to have a greater performance than System B if $P_T^A > P_T^B$.

As for the case of optimal configuration, this comparison will often only be possible to be performed numerically. However, by making the same assumptions as in Sect. 4.1 it is possible to achieve analytical insight into the comparison of two systems.

By using (4.3) and (4.4) we have

$$\frac{I(\boldsymbol{x}; \boldsymbol{y}^A)}{T} = \frac{I(\boldsymbol{\xi}^A; \boldsymbol{\phi}^A)}{T} \simeq \frac{(H(\boldsymbol{\xi}^A) - H(\boldsymbol{\zeta}^A))}{T} \tag{4.31}$$

and similarly

$$\frac{I(\boldsymbol{x}; \boldsymbol{y}^B)}{T} = \frac{I(\boldsymbol{\xi}^B; \boldsymbol{\phi}^B)}{T} \simeq \frac{(H(\boldsymbol{\xi}^B) - H(\boldsymbol{\zeta}^A))}{T}. \tag{4.32}$$

These two equations show that the performance will be based on the difference between two factors: $H(\boldsymbol{\xi})$ which is the entropy of source p.d.f. measured in the communications frame and $H(\boldsymbol{\zeta})$ which is a measure of the 'noisiness' of the positioning system. The $H(\boldsymbol{\xi})$ depend on the positional statistics and the

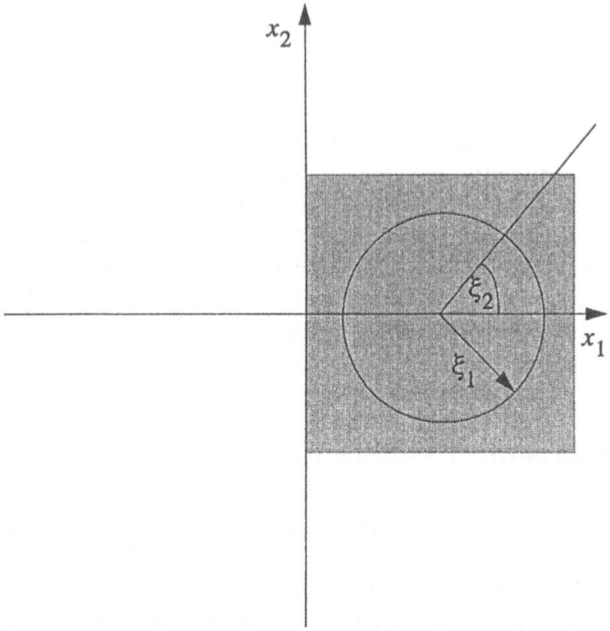

Fig. 4.1. Diagram of Polar System with Positional p.d.f.

system geometry. The $H(\zeta)$ will depend on such factors as bandwidth, array gain and signal strength.

Example 4.2

Suppose a system designer has to decide between two possible positioning systems, called A and B. The systems have to operate on a two-dimensional co-ordinate system (x_1, x_2). Each measurement is independent. The *a priori* probability density is uniform over a square with unit length sides centred at $(\frac{1}{2}, 0)$. The bottom side of the square is oriented to be parallel to the x_1 axis.

System A is a polar system. Its reference site may be placed anywhere along the x_1 axis. Denote its position as $(0, a)$. This system is shown in Fig. 4.1. System B is an angle-angle system. One reference site must be placed at $(0, 0)$. The other reference site can be placed anywhere along the positive x_1 axis. Denote its position as $(0, b)$. This system is shown in Fig. 4.2.

For both systems we make the following assumptions: the measurement errors in the communication frame are additive and independent of position, both systems have the same measurement entropy error $H(\zeta)$, both systems take the same length of time for a measurement, and the variance of the measurement error is very small.

Using (4.31) and (4.32) we can write that

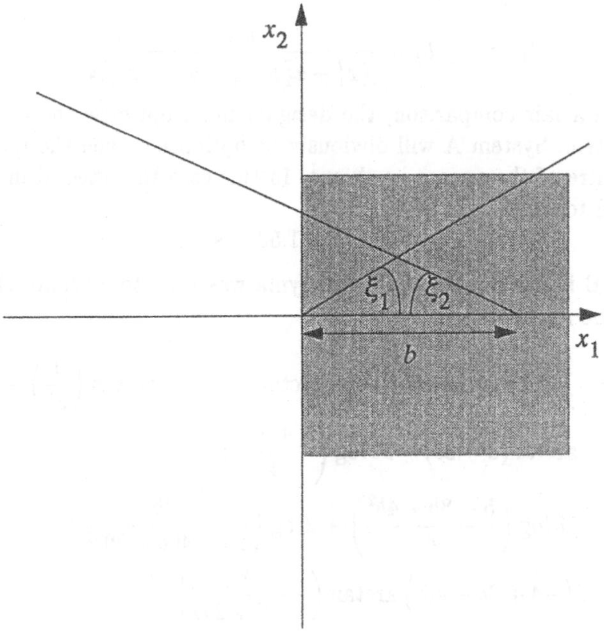

Fig. 4.2. Diagram of Angle-Angle System with Positional p.d.f.

$$\frac{I(\boldsymbol{x};\boldsymbol{y}^A)}{T} = \frac{I(\boldsymbol{\xi}^A;\boldsymbol{\phi}^A)}{T} = \frac{(H(\boldsymbol{\xi}^A) - H(\boldsymbol{\zeta}))}{T} \tag{4.33}$$

and

$$\frac{I(\boldsymbol{x};\boldsymbol{y}^B)}{T} = \frac{I(\boldsymbol{\xi}^B;\boldsymbol{\phi}^B)}{T} = \frac{(H(\boldsymbol{\xi}^B) - H(\boldsymbol{\zeta}))}{T}. \tag{4.34}$$

From these equations it is clear that the optimal configuration and comparison of the two systems can be achieved by simply considering $H(\boldsymbol{\xi}^A)$ and $H(\boldsymbol{\xi}^B)$.

By using the same reasoning as in the derivation of (4.9) we can write

$$H(\boldsymbol{\xi}^A;a) = \int_0^1 \int_{-\frac{1}{2}}^{\frac{1}{2}} dx_1\, dx_2 \log J_A(x_1,x_2;a) + \int_0^1 \int_{-\frac{1}{2}}^{\frac{1}{2}} dx_1\, dx_2 \log 1 \tag{4.35}$$

and

$$H(\boldsymbol{\xi}^B;b) = \int_0^1 \int_{-\frac{1}{2}}^{\frac{1}{2}} dx_1\, dx_2 \log J_B(x_1,x_2;b) + \int_0^1 \int_{-\frac{1}{2}}^{\frac{1}{2}} dx_1\, dx_2 \log 1 \tag{4.36}$$

where $\boldsymbol{J_A}$ and $\boldsymbol{J_B}$ are the Jacobians of the coordinate transformations. Note that because of the log 1 term the second integral will be zero in both cases.

$\boldsymbol{J_A}$ and $\boldsymbol{J_B}$ are given by

$$J_A(x_1,x_2;a) = \frac{1}{((x_1-a)^2 + x_2^2)^{\frac{1}{2}}} \tag{4.37}$$

and

$$J_B(x_1, x_2; b) = \frac{b|x_2|}{((x_1^2 + x_2^2)((x_1 - b)^2 + x_2^2))^{\frac{1}{2}}}. \tag{4.38}$$

To make a fair comparison, the designer must optimise the configuration for each system. System A will obviously be optimised when the reference site is in the centre of the square i.e. $b = \frac{1}{2}$. In this case the integral in (4.35) can be evaluated to yield

$$H(\boldsymbol{\xi}^A) \simeq 1.53\,\text{bits} \tag{4.39}$$

System B is more interesting. Macsyma was used to evaluate the integral in (4.36). The result was

$$\begin{aligned} H(\boldsymbol{\xi}^B; b) \;=\; & \frac{1}{2}\Bigg[10 - 4\arctan\left(\frac{1}{2}\right) - \arctan(2) - 4b^2\arctan\left(\frac{1}{2b}\right) - \arctan(2b) - \\ & \arctan(2-2b) - 2b\log\left(\frac{1+4b^2}{4}\right) + \\ & 2b\log\left(\frac{5-8b+4b^2}{4}\right) + 2\log\left(\frac{8b}{25-40b+20b^2}\right) + \\ & \left(-4+8b-4b^2\right)\arctan\left(-\frac{1}{-2+2b}\right)\Bigg] \end{aligned} \tag{4.40}$$

This equation is plotted in Fig. 4.3. The maximum is at $b \simeq 0.717$. This is a sensible, though not immediately obvious, result. The value of $H(\boldsymbol{\xi}^B, b)$ at the maximum is 1.79 bits.

As $H(\boldsymbol{\xi}^B)$ is greater than $H(\boldsymbol{\xi}^A)$ the designer should choose System B. Real systems will normally have errors with at least an inverse square law dependency on distance. This will make the performance more sensitive to the location of reference sites and the type of system. Of course in a real evaluation many other factors, such as cost and site feasibility, would play a role in the decision.

□

4.3 Measurement Strategy

The selection of the measurement vector M_T will often be governed by a constraint on the way the various objects must be measured. The constraints may take a number of different forms, for example:

- The system must measure the position of an entire fleet of ambulances. Each measurement must be to an accuracy of 10 metres r.m.s. and the measurement of each vehicle must be updated every 60 seconds.
- The system must monitor the position of every aeroplane within 100 kilometres to an accuracy of 100 metres.
- The system must measure the position of the first six taxis to an accuracy of 20 metres r.m.s. within 30 seconds (a taxi dispatch system).

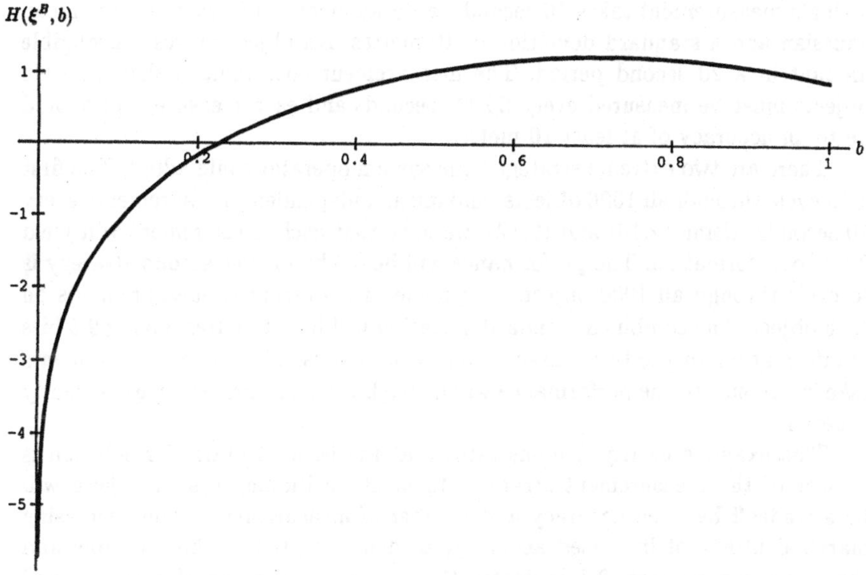

Fig. 4.3. Plot of Entropy versus b

Given particular source statistics and a measurement constraint, the system operator should choose a measurement strategy that optimises the performance of the system. Mathematically this can be stated as given a measurement vector $\boldsymbol{M}_T^A$ and a second vector $\boldsymbol{M}_T^B$, if both vectors produce measurements that satisfy the measurement constraint, then $\boldsymbol{M}_T^A$ is considered superior to $\boldsymbol{M}_T^B$ if

$$P_T(\boldsymbol{X}^A, \boldsymbol{Y}^A) > P_T(\boldsymbol{X}^B, \boldsymbol{Y}^B). \tag{4.41}$$

In certain cases there may be an optimal measurement vector. Denote this measurement vector as $\boldsymbol{M}_T^{\text{opt}}$. The performance of the system in this case will be $I(\boldsymbol{X}^{\text{opt}}; \boldsymbol{Y})/T$.

The importance of measurement strategy will vary between different types of positioning devices. Some devices are severely limited by physical factors (e.g. the continuous scanning of a radar). Whereas other systems can choose which object to measure and so have a great deal of flexibility. In any case it will be of benefit to consider the factors affecting the selection of measurement strategy.

Example 4.3

The following example shows how a simple analysis can be used to make decisions on measurement strategy and optimal resolution. Suppose a one-dimensional system has 1000 objects. Assume each object has a zero-mean gaussian p.d.f. with standard deviation of 5 120 metres. The decorrelation time for position of an object is 10 000 seconds. Measurement error is additive, and a single measurement takes 10 seconds, with an error p.d.f. which is zero-mean gaussian and a standard deviation of 10 metres. An object moves a negligible amount in a 20 second period. The measurement constraint is that all 1000 objects must be measured every 20 000 seconds and each measurement should be to an accuracy of at least 10 metres.

There are two extreme strategies the system operator could adopt. The first is to cycle through all 1000 objects, making an independent measurements every 10 seconds. Using (2.10) and (2.12), we have that each measurement will yield 9 bits of information. The performance will be 0.9 bit/s. The second strategy is to cycle through all 1000 objects, but make two consecutive measurements on each object. The combined standard deviation will be 7.1 metres, giving 9.5 bits of information for the two consecutive measurements. These two measurements take 20 seconds so the performance will be 0.5 bits/s. The first strategy is clearly superior.

This example clearly demonstrates that the issue of optimal resolution is a facet of the measurement strategy. In most positioning systems there will be a tradeoff between accuracy and number of measurements. The decreasing marginal utility of increased accuracy, as demonstrated in this example and the discussion in Sect. 2.1 indicates that the optimal value of accuracy will normally be the minimum acceptable accuracy. In many systems this will be closely related to the physical resolution of the system. The trade-off between resolution and information was first discussed by Woodward [128]. □

4.4 Independent Error Approximation

It can prove difficult to optimise practical systems using (4.1). This is because such systems make many measurements so that the optimisation problem has a very large dimensionality. As well, $I(\boldsymbol{X};\boldsymbol{Y})$ may be difficult or impossible to evaluate. This problem can be overcome if the system can satisfy the following conditions:

- The errors for each measurement are independent,
- the measurement errors are additive in the communications frame,
- the variance of the measurement errors are small compared to the variance of the positional p.d.f,
- the region of interest is all of space [1] (limits of integrals are from $-\infty$ to ∞),
- the system is well-defined and $\boldsymbol{g}^{-1} = \boldsymbol{f}$.

Many practical systems will satisfy these conditions. For such systems we can assume, to a high degree of accuracy, that $H(\boldsymbol{Y}) \simeq H(\boldsymbol{X})$, so we have that

[1] This does not prevent the positional p.d.f. from excluding certain regions.

$$I(\boldsymbol{X};\boldsymbol{Y}) \simeq H(\boldsymbol{X}) - H(\boldsymbol{Y}|\boldsymbol{X}). \tag{4.42}$$

The assumption of independent errors implies

$$H(\boldsymbol{Y}|\boldsymbol{X}) = \sum_{i=1}^{K} H(\boldsymbol{y}_i|\boldsymbol{x}_i). \tag{4.43}$$

$H(\boldsymbol{y}_i|\boldsymbol{x}_i)$ is given by

$$H(\boldsymbol{y}|\boldsymbol{x}) = \int_{-\infty}^{\infty}\int_{-\infty}^{\infty} \mathrm{d}\boldsymbol{x}\ \mathrm{d}\boldsymbol{y}\, p(\boldsymbol{x},\boldsymbol{y}) \log[p_{y|x}(\boldsymbol{y}|\boldsymbol{x})]. \tag{4.44}$$

For notational convenience the subscript i has been omitted, it will be included only when this convention causes ambiguity. Now consider

$$p_{y|x}(\boldsymbol{y}|\boldsymbol{x}) = p_{\phi|\xi}(\boldsymbol{\phi}|\boldsymbol{\xi})J(\boldsymbol{y}) \tag{4.45}$$

which, given our assumption concerning the additivity of errors in the communications frame, can be written as

$$p_{y|x}(\boldsymbol{y}|\boldsymbol{x}) = p_{\phi|\xi}(\boldsymbol{\phi}-\boldsymbol{\xi},\boldsymbol{\xi})J(\boldsymbol{y}). \tag{4.46}$$

For example a one-dimensional radial system which has gaussian distributed errors with a standard error proportional to the inverse square law could have a conditional p.d.f. of the form

$$p_{\phi|\xi}(\phi|\xi) = \frac{1}{\alpha|\xi|\sqrt{2\pi}} e^{-\frac{1}{2}\frac{(\phi-\xi)^2}{\alpha^2\xi^4}} = p_{\phi|\xi}(\phi-\xi,\xi). \tag{4.47}$$

Let $\delta\boldsymbol{\xi} = \boldsymbol{\phi} - \boldsymbol{\xi}$ and $\delta\boldsymbol{x} = \boldsymbol{y} - \boldsymbol{x}$, then

$$\delta\boldsymbol{x} = \boldsymbol{g}^{-1}(\boldsymbol{\xi}+\delta\boldsymbol{\xi}) - \boldsymbol{g}^{-1}(\boldsymbol{\xi}), \tag{4.48}$$

so that Taylor-expanding about $\boldsymbol{\xi}$ gives

$$\delta\boldsymbol{x} = \boldsymbol{J}_m(\boldsymbol{\xi})\delta\boldsymbol{\xi} + \text{Higher Order Terms} \tag{4.49}$$

where $\boldsymbol{J}_m$ is the Jacobian matrix, evaluated at $\boldsymbol{\xi}$. The Jacobian matrix for the ith measurement is defined as

$$[\boldsymbol{J}_m(\boldsymbol{\xi}_i)]_{jk} = \frac{\partial g_j^{-1}(\boldsymbol{\xi}_i)}{\partial \xi_{ki}}. \tag{4.50}$$

The small error approximation allows us to ignore the higher order terms, so we can write

$$\delta\boldsymbol{\xi} \simeq \boldsymbol{J}_m^{-1}(\boldsymbol{g}(\boldsymbol{x}))\delta\boldsymbol{x} \tag{4.51}$$

or

$$\boldsymbol{\phi} - \boldsymbol{\xi} \simeq \boldsymbol{J}_m^{-1}(\boldsymbol{g}(\boldsymbol{x}))(\boldsymbol{y}-\boldsymbol{x}), \tag{4.52}$$

and that

$$J(\boldsymbol{y}) \simeq \det(\boldsymbol{J}_m^{-1}(\boldsymbol{g}(\boldsymbol{x}))) = J(\boldsymbol{x}). \tag{4.53}$$

The forgoing assumes that $\boldsymbol{J}_m$ is not singular. Practical systems will have at most a few isolated singularities, often at the reference sites. Physical considerations (e.g. receiver overload and physical bulk) mean that objects normally cannot be placed at these singular points.

Substituting equations (4.52) and (4.53) into (4.46) gives

$$p_{y|x}(\boldsymbol{y}|\boldsymbol{x}) = p_{\phi|\xi}(\,\boldsymbol{J}_m^{-1}(\boldsymbol{g}(\boldsymbol{x}))(\boldsymbol{y}-\boldsymbol{x}), \boldsymbol{g}(\boldsymbol{x})\,)J(\boldsymbol{x}). \tag{4.54}$$

Putting this equation into (4.44) gives

$$\begin{aligned} H(\boldsymbol{y}|\boldsymbol{x}) &= -\int_{-\infty}^{\infty} \mathrm{d}\boldsymbol{x}\, p_x(\boldsymbol{x}) \int_{-\infty}^{\infty} \mathrm{d}\boldsymbol{y}\, p_{\phi|\xi}(\,\boldsymbol{J}_m^{-1}(\boldsymbol{g}(\boldsymbol{x}))(\boldsymbol{y}-\boldsymbol{x}), \boldsymbol{g}(\boldsymbol{x})\,)J(\boldsymbol{x}) \\ &\quad \cdot \log[p_{\phi|\xi}(\,\boldsymbol{J}_m^{-1}(\boldsymbol{g}(\boldsymbol{x}))(\boldsymbol{y}-\boldsymbol{x}), \boldsymbol{g}(\boldsymbol{x})\,)J(\boldsymbol{x})] \end{aligned} \tag{4.55}$$

where the joint p.d.f. $p(\boldsymbol{x}, \boldsymbol{y})$ has been separated using the joint probability theorem.

Let us define

$$Q_{\xi|\phi}(\boldsymbol{x}) = -\int_{-\infty}^{\infty} \mathrm{d}\boldsymbol{y}\, p_{\phi|\xi}(\,\boldsymbol{y}-\boldsymbol{x}, \boldsymbol{g}(\boldsymbol{x})\,) \log[p_{\phi|\xi}(\,\boldsymbol{y}-\boldsymbol{x}, \boldsymbol{g}(\boldsymbol{x})\,)] \tag{4.56}$$

and

$$\boldsymbol{J}_w(\boldsymbol{x}) = \boldsymbol{J}_m^{-1}(\boldsymbol{g}(\boldsymbol{x})). \tag{4.57}$$

The quantity $Q_{\phi|\xi}$ represents the entropy of the conditional error in the communications frame. The matrix $\boldsymbol{J}_w(\boldsymbol{x})$ is the Jacobian matrix of the transformation of $\boldsymbol{\xi} \longrightarrow \boldsymbol{x}$, so that

$$|\det[\boldsymbol{J}_w(\boldsymbol{x})]| = J(\boldsymbol{x}). \tag{4.58}$$

Using these definitions, (4.55) becomes

$$\begin{aligned} H(\boldsymbol{y}|\boldsymbol{x}) &= -\int_{-\infty}^{\infty} \mathrm{d}\boldsymbol{x}\, p_x(\boldsymbol{x}) \int_{-\infty}^{\infty} \mathrm{d}\boldsymbol{y}\, p_{\phi|\xi}(\,\boldsymbol{J}_w(\boldsymbol{x})(\boldsymbol{y}-\boldsymbol{x}), \boldsymbol{g}(\boldsymbol{x})\,)J(\boldsymbol{x}) \\ &\quad \cdot \log[p_{\phi|\xi}(\,\boldsymbol{J}_w(\boldsymbol{x})(\boldsymbol{y}-\boldsymbol{x}), \boldsymbol{g}(\boldsymbol{x})\,)J(\boldsymbol{x})], \end{aligned} \tag{4.59}$$

or making the change of variable, of the form

$$\boldsymbol{u} = \boldsymbol{J}_w(\boldsymbol{x})(\boldsymbol{y}-\boldsymbol{x}) \tag{4.60}$$

we have

$$\begin{aligned} H(\boldsymbol{y}|\boldsymbol{x}) &= -\int_{-\infty}^{\infty} \mathrm{d}\boldsymbol{x}\, p_x(\boldsymbol{x}) \int_{-\infty}^{\infty} \mathrm{d}\boldsymbol{u}\, p_{\phi|\xi}(\,\boldsymbol{u}, \boldsymbol{g}(\boldsymbol{x})\,)J(\boldsymbol{x})|\det \boldsymbol{J}_w^{-1}(\boldsymbol{x})| \\ &\quad \cdot \log[p_{\phi|\xi}(\,\boldsymbol{u}, \boldsymbol{g}(\boldsymbol{x})\,)J(\boldsymbol{x})]. \end{aligned} \tag{4.61}$$

By use of (4.56) and (4.58), Eqn. (4.61) becomes

$$H(\boldsymbol{y}|\boldsymbol{x}) = \int_{-\infty}^{\infty} \mathrm{d}\boldsymbol{x}\, p_x(\boldsymbol{x}) Q_{\phi|xi}(\boldsymbol{x}) - \int_{-\infty}^{\infty} \mathrm{d}\boldsymbol{x}\, p_x(\boldsymbol{x}) \log[J(\boldsymbol{x})] \int_{-\infty}^{\infty} \mathrm{d}\boldsymbol{u}\, p_{\phi|\xi}(\boldsymbol{u}, \boldsymbol{g}(\boldsymbol{x})) \tag{4.62}$$

and, remembering that p.d.fs are normalised to 1, gives

$$H(\boldsymbol{y}|\boldsymbol{x}) = \int_{-\infty}^{\infty} \mathrm{d}\boldsymbol{x}\, p_x(\boldsymbol{x})(Q_{\phi|\boldsymbol{x}i}(\boldsymbol{x}) - \log[J(\boldsymbol{x})]). \tag{4.63}$$

Substituting this equation into (4.42) and (4.43) yields

$$I(\boldsymbol{X};\boldsymbol{Y}) \simeq H(\boldsymbol{X}) - \sum_{i=1}^{K} \int_{-\infty}^{\infty} \mathrm{d}\boldsymbol{x}_i\, p_{x_i}(\boldsymbol{x})(Q_{\phi|\xi}(\boldsymbol{x}_i) - \log[J(\boldsymbol{x}_i)]). \tag{4.64}$$

This result will be called the independent error approximation. If $Q_{\phi|\xi}$ is independent of $\boldsymbol{x}$ this result reduces to that of Sect. 4.1.

Example 4.4

Consider a system where the positional p.d.f. of each measurement is identical. Suppose further the errors are gaussian distributed, but each can be a function position. Then $p_{\phi|\xi}$ can be written in the form

$$p_{\phi|\xi}(\boldsymbol{\phi} - \boldsymbol{\xi}, \boldsymbol{\xi}) = \frac{\det(\boldsymbol{C}(\boldsymbol{\xi}))^{\frac{1}{2}}}{(2\pi)^{\frac{n}{2}}} \exp\left(-\frac{1}{2}\delta\boldsymbol{\xi}^{\mathrm{T}}\boldsymbol{C}(\boldsymbol{\xi})\delta\boldsymbol{\xi}\right), \tag{4.65}$$

where $\boldsymbol{C}^{-1}(\boldsymbol{\xi})$ is the covariance matrix.

Accordingly, using Eqn. (6.15.16) of Jones [53], the entropy can be shown to be

$$Q_{\phi|\xi}(\boldsymbol{x}) = \log\left[\frac{(2\pi e)^{\frac{n}{2}}}{\det(\boldsymbol{C}(\boldsymbol{g}(\boldsymbol{x})))^{\frac{1}{2}}}\right]. \tag{4.66}$$

If K measurements are made, then the mutual information is given by

$$I(\boldsymbol{X};\boldsymbol{Y}) \simeq H(\boldsymbol{X}) - K\int_{-\infty}^{\infty} \mathrm{d}\boldsymbol{x}\, p_x(\boldsymbol{x})\left(\log\left[\frac{(2\pi e)^{\frac{n}{2}}}{\det(\boldsymbol{C}(\boldsymbol{g}(\boldsymbol{x})))^{\frac{1}{2}}J(\boldsymbol{x})}\right]\right). \tag{4.67}$$

This equation can be evaluated numerically, and in some cases analytically. A method of calculating $\boldsymbol{C}$ will be discussed in Chap. 6 (see (6.57)). □

Example 4.5

Consider a two-dimensional radial-radial positioning system. Suppose the system satisfies all the assumptions outlined at the beginning of this section. As well suppose the errors at each of the measurement sites are independent and normally distributed, with a variance with an inverse square law proportionality. Each measurement has the same *a priori* p.d.f.

If the reference sites are located at $(0,0)$ and $(b,0)$, then the Jacobian will be given by

$$J(x_1, x_2) = \frac{|x_2|b}{\sqrt{(x_1^2 + x_2^2)((x_1 - b)^2 + x_2^2)}}. \tag{4.68}$$

For this system the matrix C will be given by

$$C(x_1,x_2) = \begin{pmatrix} \frac{1}{\alpha_1(x_1^2+x_2^2)} & 0 \\ 0 & \frac{1}{\alpha_2((x_1-b)^2+x_2^2)} \end{pmatrix} \tag{4.69}$$

where α_1 and α_2 are constants of proportionality.

Substitution of these two equations into (4.67) gives

$$\begin{aligned} I(X;Y) \simeq\ & H(X) - \\ & K \int_{-\infty}^{\infty}\int_{-\infty}^{\infty} \mathrm{d}x_1\, \mathrm{d}x_2\, p_x(x_1,x_2) \log\left[\frac{2\pi e\alpha_1\alpha_2(x_1^2+x_2^2)((x_1-b)^2+x_2^2)}{|x_2|b}\right]. \end{aligned} \tag{4.70}$$

This equation is the same as for an angle-angle system with constant errors (see (4.38)). Accordingly the optimisation shown in Example 4.2 would also apply to this system if the positional p.d.f. were changed to a uniform distribution bounded by the lines $x_1 = 0, x_1 = 1, x_2 = 0, x_2 = \frac{1}{2}$, i.e. the p.d.f. were folded about the x_1-axis to avoid ambiguity. □

4.5 Classes of Devices

Section 4.2 showed how it was possible to compare two different types of systems. This process would be used by a systems engineer to help make a selection decision. In order to gain an understanding of the general properties of positioning devices it is also of interest to compare the performance of classes of devices.

A class of systems will be defined in the following manner: Let $\mathcal{M}$ be a class of functionals. Let $\mathcal{T}$ be a second class of functionals. Suppose both these sets have the same index, α. The index may be discrete or continuous. Remember from Sect. 2.2, that a positioning device $\boldsymbol{S}$ is defined by its mappings $(\boldsymbol{g},\boldsymbol{q},\boldsymbol{f})$. Then the class of devices $\mathcal{S}$ is the set

$$\begin{aligned} \mathcal{S} = \{&\boldsymbol{S} : \boldsymbol{S}_\alpha = (\boldsymbol{g}_\alpha,\boldsymbol{q}_\alpha,\boldsymbol{g}_\alpha^{-1}) = (\boldsymbol{M}_\alpha(\boldsymbol{g}_g),\boldsymbol{T}_\alpha(\boldsymbol{q}_g),\boldsymbol{M}_\alpha(\boldsymbol{g}_g^{-1})); \\ &\boldsymbol{M}_\alpha \in \mathcal{M}; \boldsymbol{T}_\alpha \in \mathcal{T}\} \end{aligned} \tag{4.71}$$

where the positioning device $\boldsymbol{S}_g = (\boldsymbol{g}_g,\boldsymbol{q}_g,\boldsymbol{g}_g^{-1})$ is called the generator of the class. From this definition we see that a class of devices is defined by $(\mathcal{M},\mathcal{T},\mathcal{S})$. Of course for such a class to have practical interest, each member of the class should be physically realizable.

In order to make a positioning system, the positioning device has to make measurements on a set of objects with source statistics $\beta(t)$ using a measurement vector $\boldsymbol{M}_T$. This means that $\boldsymbol{S}$ operates on a positional p.d.f. $p_X(\boldsymbol{X})$.

For each member $\boldsymbol{S}_\alpha$ of a class of devices, it is possible to calculate $I(\boldsymbol{X};\boldsymbol{Y}_\alpha)$. Two members of the class, $\boldsymbol{S}_{\alpha'}$ and $\boldsymbol{S}_\alpha$, will be considered equivalent if $I(\boldsymbol{X};\boldsymbol{Y}_{\alpha'}) = I(\boldsymbol{X};\boldsymbol{Y}_\alpha)$ for every possible $p_X(\boldsymbol{X})$. The members of a class that are equivalent form an equivalence subset of $\boldsymbol{S}$. An example of an equivalence class is if $\boldsymbol{M}_{\alpha'}$ and $\boldsymbol{T}_{\alpha'}$ can be derived from $\boldsymbol{M}_\alpha$ and $\boldsymbol{T}_\alpha$ by consistently relabelling the reference sites. This is because the mutual information cannot be changed by a simple relabelling.

A class is called *trivial* if all members of the class belong to the same equivalence set. The proposition set out below describes an important case when a class of devices is trivial.

Proposition 4.6 *A class of systems is trivial, if for all* α, $\boldsymbol{q}_\alpha = \boldsymbol{g}_\alpha \boldsymbol{g}_g^{-1} \boldsymbol{q}_g \boldsymbol{g}_g \boldsymbol{g}_\alpha^{-1}$.

Proof

We have that

$$\boldsymbol{y} = \boldsymbol{g}^{-1}(\boldsymbol{q}(\boldsymbol{g}(\boldsymbol{x}))) \tag{4.72}$$

but

$$\boldsymbol{y}_\alpha = \boldsymbol{g}_\alpha^{-1}(\boldsymbol{q}_\alpha(\boldsymbol{g}_\alpha(\boldsymbol{x}))), \tag{4.73}$$

which gives

$$\boldsymbol{y}_\alpha = \boldsymbol{g}_\alpha^{-1}(\boldsymbol{g}_\alpha(\boldsymbol{g}_g^{-1}(\boldsymbol{q}_g(\boldsymbol{g}_g(\boldsymbol{g}_\alpha^{-1}(\boldsymbol{g}_\alpha(\boldsymbol{x}))))))), \tag{4.74}$$

or

$$\boldsymbol{y}_\alpha = \boldsymbol{g}_g^{-1}(\boldsymbol{q}_g(\boldsymbol{g}_g(\boldsymbol{x}))). \tag{4.75}$$

Given this is true for every $\boldsymbol{x}$ and for each α, it follows that $I(\boldsymbol{X};\boldsymbol{Y}_\alpha) = I(\boldsymbol{X};\boldsymbol{Y})$ for every possible positional p.d.f. and for every member of the set.

End of Proof

This proposition simply states that no information is either gained or lost by a consistent 1:1 transformation of the coordinate system. For example GPS can be thought of as using a hyperbolic-hyperbolic-hyperbolic co-ordinate system (i.e. using three independent time-difference-of-arrival measurements) or a pseudorange-pseudorange-pseudorange- pseudorange co-ordinate system. However to move from one co-ordinate system to the other, $\boldsymbol{q}_\alpha$ must be transformed as shown in the above proposition, so the average mutual information will be the same for either co-ordinate system.

Consider a class of devices, $\boldsymbol{\mathcal{S}}$, operating on a positional p.d.f. $p_X(\boldsymbol{X})$. A device $\boldsymbol{S}_{\alpha'}$ is called the optimum device w.r.t $p_X(\boldsymbol{X})$ if $I(\boldsymbol{X};\boldsymbol{Y}_{\alpha'}) > I(\boldsymbol{X};\boldsymbol{Y}_\alpha)$ for all $\boldsymbol{\alpha}$ not equal to $\boldsymbol{\alpha}'$. If $\boldsymbol{S}_{\alpha'}$ is the optimum device for all possible positional p.d.fs then it will be called the universally optimum device of the class. If the optimum device is a member of an equivalence subset, then each member of the subset will trivially be the optimum device.

These definitions should allow general theorems to be proved about classes of systems. This should assist in the selection of particular systems and aid in the invention of new systems.

Example 4.7

Consider a class of two-dimensional devices of which satisfy the independent error approximation and have gaussian distributed errors. The generator is

$$g_1(\boldsymbol{x}) = \sqrt{(\boldsymbol{x}-\boldsymbol{u}_1)^{\mathrm{T}}(\boldsymbol{x}-\boldsymbol{u}_1)} = \boldsymbol{r}_1, \tag{4.76}$$

$$g_2(\boldsymbol{x}) = \sqrt{(\boldsymbol{x} - \boldsymbol{u}_2)^{\mathrm{T}}(\boldsymbol{x} - \boldsymbol{u}_2)} = \boldsymbol{r}_2, \tag{4.77}$$

where r_1 and r_2 are the euclidean distance from the reference sites located at $\boldsymbol{u}_1$ and $\boldsymbol{u}_2$ respectively.

Because the generator satisfies the small error approximation, q_g is completely determined by a definition of A_g, the inverse of the correlation matrix. Suppose A_g is defined as

$$A_g = \begin{pmatrix} \frac{1}{\sigma^2(r_1)} & 0 \\ 0 & \frac{1}{\sigma^2(r_2)} \end{pmatrix}. \tag{4.78}$$

This is the matrix corresponding to a system that has a error variance that depends upon distance ($\sigma(r)$) and takes independent measurements from two different reference sites. In order to fully specify the class we must define $\boldsymbol{M}_\alpha$ and $\boldsymbol{T}_\alpha$ by outlining the method of measurement. Let us call a single measurement of distance between an object and reference site, or between two reference sites a *leg*. Consider the class of systems that make two independent measurements, each measurement consisting of two legs.

Then $\boldsymbol{M}_\alpha$ is defined as

$$\boldsymbol{g}_\alpha(\boldsymbol{x}) = \boldsymbol{G}(\alpha)\boldsymbol{g}(\boldsymbol{x}) \tag{4.79}$$

where $\boldsymbol{G}(\alpha)$ is an affine transformation with the following properties:

- the sum of the absolute value of the elements of each row should be equal to 2 (i.e. each measurement uses two legs),
- the determinant of $\boldsymbol{G}(\alpha)$ must be non-zero (for independent measurements)
- each element of $\boldsymbol{G}(\alpha)$ can only be 0, 1 or 2, and
- the determinant of $\boldsymbol{G}(\alpha)$ must be positive.

The last two constraints reduce the size of the class by eliminating some systems which are physically identical to other members of the class.

If $\boldsymbol{J}_{\boldsymbol{M}_g}(\boldsymbol{x})$ is the Jacobian matrix of the generating function then the Jacobian matrix of $\boldsymbol{g}_\alpha$ is given by

$$\boldsymbol{J}_{\boldsymbol{M}_\alpha} = \boldsymbol{G}(\alpha)\boldsymbol{J}_{\boldsymbol{M}_g}(\boldsymbol{x}). \tag{4.80}$$

Because the measurement errors are assumed to be gaussian for this class, $\boldsymbol{T}_\alpha$ can be defined in terms of $\boldsymbol{A}_\alpha$. The independence of the measurements implies that $\boldsymbol{A}_\alpha$ is defined as follows:

$$\boldsymbol{A}_\alpha = \begin{pmatrix} \frac{1}{\sigma^2(r_1)|G(\alpha)_{11}|+\sigma^2(r_2)|G(\alpha)_{12}|} & 0 \\ 0 & \frac{1}{\sigma^2(r_1)|G(\alpha)_{21}|+\sigma^2(r_2)|G(\alpha)_{22}|} \end{pmatrix}. \tag{4.81}$$

This is a finite class with 13 members. The index α can be an integer, the members of the class being numbered arbitrarily from 1 to 13. In Appendix G the class is enumerated, described and representative loci plotted. The loci of this class can be divided into four subsets: radial-radial (one member), radial-elliptic (four members), radial- hyperbolic (four members) and elliptic-hyperbolic (four members). Note that these are not equivalence subsets.

It should be emphasized that each of the devices belonging to this class are physically realizable positioning devices. For example, one realization would be for the devices to be self positioning and use pulses to measure distance, with a transponder at each reference site. One member of the class would be

$$G(\alpha_{rh}) = \begin{pmatrix} 1 & -1 \\ 2 & 0 \end{pmatrix}. \tag{4.82}$$

This device measures the time difference of arrival of two pulses ($r_1 - r_2$) and the round trip time for a pulse to go from the object to the first reference site and back again ($2r_1$). Accordingly it is a hyperbolic-radial device. The definition of (4.81) is simply a statement that if a transponder is used to retransmit the pulse in each case then the variance of the measurement should be equal to the sum of the variances of each individual leg. The requirement that the two measurements are independent (i.e. the matrix is diagonal) is a reasonable design goal. Accordingly this is a physically realizable system. Each member of the class can be checked in this way.

For this class it is not difficult to calculate the optimum member by simple enumeration. First consider (see (4.67)) the expression

$$D = \log[\det(\boldsymbol{A}_\alpha)^{\frac{1}{2}}|\det(\boldsymbol{G}(\alpha)\boldsymbol{J}_{M_s}(x))|]. \tag{4.83}$$

Let us evaluate this term for each equivalence subset. The equivalence sets and sufficient information to evaluate D are itemised below. Note that radial(r_1) means radial distance measured from the first reference site, while radial(r_2) means measured from the second reference site.

Radial-Radial Equivalence Subset

- Number of members in subset = 1
- $\det(\boldsymbol{A}_\alpha)^{\frac{1}{2}} = \frac{1}{4\sigma(r_1)^2\sigma(r_2)^2}$
- $|\det(G(\alpha))| = 4$
- $D = -\log[\sigma(r_1)^2\sigma(r_2)^2] + \log[|\det\boldsymbol{J}_{M_s}(x)|]$

Radial(r_1)-Elliptic and Radial(r_1)-Hyperbolic Equivalence Subset

- Number of members in subset = 4
- $\det(\boldsymbol{A}_\alpha)^{\frac{1}{2}} = \frac{1}{2(\sigma(r_1)^2+\sigma(r_2)^2)\sigma(r_1)^2}$
- $\det(G(\alpha)) = 2$

- $D = -\log[(\sigma(r_1)^2 + \sigma(r_2)^2)\sigma(r_1)^2] + \log[|\det \boldsymbol{J}_{M_g}(x)|]$

Radial(r_2)-Elliptic and Radial(r_1)-Hyperbolic Equivalence Subset

- Number of members in subset = 4
- $\det(\boldsymbol{A}_\alpha)^{\frac{1}{2}} = \frac{1}{2(\sigma(r_1)^2+\sigma(r_2)^2)\sigma(r_2)^2}$
- $\det(G(\alpha)) = 2$
- $D = -\log[(\sigma(r_1)^2 + \sigma(r_2)^2)\sigma(r_2)^2] + \log[|\det \boldsymbol{J}_{M_g}(x)|]$

Elliptic-Hyperbolic Equivalence Subset

- Number of members in subset = 4
- $\det(\boldsymbol{A}_\alpha)^{\frac{1}{2}} = \frac{1}{(\sigma(r_1)^2+\sigma(r_2)^2)^2}$
- $\det(G(\alpha)) = 2$
- $D = -\log[\frac{1}{2}(\sigma(r_1)^2 + \sigma(r_2)^2)^2] + \log[|\det \boldsymbol{J}_{M_g}(x)|]$

Now we have that

$$\begin{aligned}
\sigma(r_1)^2\sigma(r_2)^2 &\leq (\sigma(r_1)^2 + \sigma(r_2)^2)\sigma(r_1)^2, \\
\sigma(r_1)^2\sigma(r_2)^2 &\leq (\sigma(r_1)^2 + \sigma(r_2)^2)\sigma(r_2)^2, \\
\sigma(r_1)^2\sigma(r_2)^2 &\leq \frac{1}{2}(\sigma(r_1)^2 + \sigma(r_2)^2)^2,
\end{aligned}$$

so the radial-radial equivalence subset will have the maximum value for D, no matter what the value of r_1 and r_2, independently of the position of the object. Accordingly, by examining (4.67) we see that the radial-radial equivalence subset will be optimal no matter what form the positional p.d.f. assumes: the subset is universally optimal. This method can be extended to higher-dimensional transit systems and systems with different numbers of legs.

Note this example demonstrates that devices with different loci can belong to the same equivalence set. It is also worth observing that one possible definition for $A(\alpha)$ would be

$$A(\alpha) = G^{\mathrm{T}}(\alpha)A_g G(\alpha). \tag{4.84}$$

However a substitution of (4.80) and (4.84) into (4.67) shows that each member of the class would have the same average mutual information: the class would be trivial. This is in accord with Proposition 4.6.

□

This is an important example as it shows it is possible to obtain general results without having to specify the *a priori* distribution. This removes an objection that is sometimes levelled at bayesian type theories which need to assume an *a priori* distribution [99]. Here we assume that there will be an *a priori* distribution, but are able to obtain results independent of the form of the distribution.

The following example shows how it is possible to define a class of systems with a continuous index.

Example 4.8

Consider a two-dimensional class of devices which satisfy the small error approximation and have gaussian distributed errors. Define $\boldsymbol{h}(\boldsymbol{x};\alpha)$ as

$$h_1(\boldsymbol{x};\alpha) = \sqrt{\boldsymbol{x}^{\mathrm{T}}\boldsymbol{x}} \tag{4.85}$$

$$h_2(\boldsymbol{x},\alpha) = \sqrt{(\boldsymbol{x}-\boldsymbol{u}_2)^{\mathrm{T}}(\boldsymbol{x}-\boldsymbol{u}_2)} \tag{4.86}$$

where $\boldsymbol{u}_2 = (0,\alpha)$.

This corresponds to a radial-radial positioning system with one reference site at $(0,0)$ and one at $(0,\alpha)$.

Let the generator of the system be $(\boldsymbol{h}(\boldsymbol{x};1), \boldsymbol{q}_g, \boldsymbol{h}^{-1}(\boldsymbol{x},1))$, where $\boldsymbol{q}_g$ is additive gaussian with correlation matrix given by

$$\boldsymbol{A}_g = \begin{pmatrix} \frac{\beta}{h_1(\boldsymbol{x})^2} & 0 \\ 0 & \frac{\beta}{h_2(\boldsymbol{x}_2;1)^2} \end{pmatrix} \tag{4.87}$$

with β being a constant of proportionality.

$\boldsymbol{M}_\alpha$ is defined as

$$g_\alpha(\boldsymbol{x}) = \boldsymbol{h}(\boldsymbol{x};\alpha) \tag{4.88}$$

and $\boldsymbol{T}_\alpha$ is defined by the correlation matrix

$$\boldsymbol{A}(\alpha) = \begin{pmatrix} \frac{\beta}{h_1(\boldsymbol{x})^2} & 0 \\ 0 & \frac{\beta}{h_2(\boldsymbol{x};\alpha)^2} \end{pmatrix}. \tag{4.89}$$

This class of devices is simply the class of radial-radial devices with one reference site at $(0,0)$ and the other at $(0,\alpha)$, with additive gaussian errors whose variance obeys the inverse square law. There will be no universal optimum, the optimum device will depend on the form of $p_X(\boldsymbol{X})$. There is a universally worst device, i.e. $\alpha = 0$.

If $p_X(\boldsymbol{X})$ is chosen as to be same as the positional p.d.f. in Example 4.5, then the optimum with respect to this p.d.f. will be the same as in Example 4.5. □

5. Coding

Chapter 2 showed that coding for a positioning system can be divided into three processes: the source, waveform coding and physical coding. In Sect. 5.1 of this chapter a method of calculating the source information rate is derived. Waveform coding is briefly discussed in Sect. 5.2. In Sect. 5.3, there is a classification of the various methods of physical coding.

5.1 The Source

The source of information in a positioning system is the movement of the object or objects being measured. The source is modelled here as a stochastic vector process. As in any communications system it is important to establish the amount of information that the source is producing. This can provide a first estimate of the required capacity of the system.

One possible measure for the source information rate would be the entropy $H(\boldsymbol{X})$, but this measure is unsuitable because the entropy is not invariant and so will differ in the world and communications frames. The reason for this is that the entropy as defined here is really a *differential entropy.* The true entropy for a continuous process is infinite. This is a consequence of the fact that measuring a continuous process to infinite accuracy requires infinite information transfer. Accordingly, a meaningful definition of the rate of information produced by a moving object will include a specification of the accuracy of the measurement. This is different to a typical communications system where the source produces information at a rate independent of the measurement process. We can now understand what the capacity of a positioning system (see Chap. 3) means: a capacity limit provides a lower bound on the accuracy available from a positioning system.

Our approach is to use Shannon's definition of source information rate [103, 102]. Analyses based on this definition are often called rate distortion theory [8]. The scalar version of this theory has been widely used [81, 6, 91, 51, 101, 7, 126],

Here we will use an approach used by Berger [8] generalised to vector stochastic processes. This generalisation is achieved in the following manner.

Consider a stochastic vector process $\boldsymbol{\beta}(t)$, where K measurements are made at intervals determined by the measurement vector $\boldsymbol{M}_T$. This generates a vector of positions $\boldsymbol{X} = (\boldsymbol{x}_1, \boldsymbol{x}_2, \ldots, \boldsymbol{x}_K)$. Suppose that measurements are made on this vector, generating a vector $\boldsymbol{Y} = (y_1, y_2, \ldots, y_K)$. Let $\boldsymbol{M}_\infty$ denote the limit of $\boldsymbol{M}_T$ as K goes to infinity. The accuracy of the measurement will be judged using a distortion measure or metric $\rho(\boldsymbol{x}, \boldsymbol{y})$. Consider all possible probability densities $q(\boldsymbol{Y}|\boldsymbol{X})$ for reproducing the position vector by $\boldsymbol{Y}$. The rate distortion function $R(D)$ of $\boldsymbol{\beta}(t)$ w.r.t. $\boldsymbol{M}_\infty$, is defined by

$$R(D) = \lim_{K\to\infty} R_K(D) \tag{5.1}$$

where

$$R_K(D) = K^{-1} \inf_{q \in Q_D} I(q) \tag{5.2}$$

and "inf" denotes the infinum.

$$Q_D = q(\boldsymbol{Y}|\boldsymbol{X}) : d(q) \equiv \int\int \mathrm{d}\boldsymbol{X}\ \mathrm{d}\boldsymbol{Y}\ p_X(\boldsymbol{X}) q(\boldsymbol{Y}|\boldsymbol{X}) \rho_K \leq D \tag{5.3}$$

$$\rho_K = \boldsymbol{K}^{-1} \sum_{i=1}^{K} \rho(\boldsymbol{x}_i, \boldsymbol{y}_i), \tag{5.4}$$

$$I(q) = I(\boldsymbol{X}; \boldsymbol{Y}) = \int\int \mathrm{d}\boldsymbol{X}\ \mathrm{d}\boldsymbol{Y}\ p_X(\boldsymbol{X}) q(\boldsymbol{Y}|\boldsymbol{X}) \log \frac{q(\boldsymbol{Y}|\boldsymbol{X})}{q(\boldsymbol{Y})} \tag{5.5}$$

and

$$q(\boldsymbol{Y}) = \int \mathrm{d}\boldsymbol{X}\ p_X(\boldsymbol{X}) q(\boldsymbol{Y}|\boldsymbol{X}). \tag{5.6}$$

We are often interested in the source information rate of $\boldsymbol{\beta}(t)$, i.e. the rate distortion function with continuous 'sampling'. In this case we have

$$R(D) = \lim_{T\to\infty} R_T(D) \tag{5.7}$$

where $R_T(D)$ is the least possible performance between $\{\boldsymbol{x}(t); 0 \leq t < T\}$ and any measurement thereof that achieves an accuracy D. The distortion between $\{\boldsymbol{x}(t); 0 \leq t < T\}$ and the $\{\boldsymbol{y}(t); 0 \leq t < T\}$ is measured by

$$\rho_T(\boldsymbol{X}, \boldsymbol{Y}) = \frac{1}{T} \int_0^T \mathrm{d}t\ \rho(\boldsymbol{x}(t), \boldsymbol{y}(t)). \tag{5.8}$$

The existence of the limit in the definition of the R(D) is guaranteed for both the discrete and continuous cases, if the source is stationary. If the source is not stationary, the limit may not exist and the definition can be meaningless.

The importance of the source information rate is justified by the following proposition.

Proposition 5.1 *Suppose the source information rate for a particular accuracy D_ρ is R. Here the subscript ρ refers to the type of distortion metric that is being used. If the channel capacity is C then it is possible to measure the position of the source with an accuracy arbitrarily close to D_ρ provided $R \leq C$. If $R > C$ then this is not possible.*

The proof of this proposition follows directly from a simple generalisation of Shannon's Theorem 21 [103]. Of course, the important issue is whether it is possible to implement a coding scheme which achieves the desired distortion. This is particularly the case for wave-based systems, where, unlike conventional communications systems, the basic coding technique is fixed by the physics of the situation. However, there are a number of performance enhancement techniques that have achieved significant improvements for such systems. These are discussed at the end of Chap. 6.

In general the rate distortion function will be very difficult to calculate. The function can be calculated for stationary gaussian sources, and it is possible to establish limits on the value of the rate distortion function for other types of sources. A useful distortion measure for positioning systems is the difference distortion measure of the mean-square-error type:

$$\rho(\boldsymbol{x}, \boldsymbol{y}) = \sum_{i=1}^{n}(x_i - y_i)^2, \tag{5.9}$$

because it corresponds to the mean square accuracy of the system.

5.1.1 Stationary Gaussian Vector Sources

In this section, the source information rate of a stationary gaussian vector source is calculated, first for the discrete case, then the continuous case.

Let $\boldsymbol{X}$ be a stationary vector with gaussian p.d.f. Without loss of generality, assume that $\boldsymbol{X}$ has zero mean.

The p.d.f. of K successive components can be written as

$$p_X(\boldsymbol{X}) = \frac{1}{(2\pi)^{\frac{nK}{2}} \det(\boldsymbol{\Phi}_K)^{\frac{1}{2}}} \exp\left(-\frac{1}{2}\boldsymbol{X}^{\mathrm{T}}\boldsymbol{\Phi}_K^{-1}\boldsymbol{X}\right) \tag{5.10}$$

where $\boldsymbol{\Phi}$ is the covariance matrix. Assume a mean-square-error-type distortion measure so that

$$\rho_K(\boldsymbol{X}, \boldsymbol{Y}) = \frac{1}{K}\sum_{i=1}^{K}\sum_{j=1}^{n}(x_{ij} - y_{ij})^2. \tag{5.11}$$

The covariance matrix is defined as

$$[\boldsymbol{\Phi}_K]_{(j-1)n+i,(t-1)n+s} = \mathcal{E}\{x_{ij}x_{st}\} \tag{5.12}$$

where i and s range from 1 to n; j and t range from 1 to K.

Now define an n by n matrix $\boldsymbol{R}(k)$ given by

$$[\boldsymbol{R}(k)]_{is} = \mathcal{E}\{x_{ij}x_{st}\} \tag{5.13}$$

where $k = j - t$.

Because of the assumption of stationarity we can write that

$$[\boldsymbol{\Phi}_K]_{(j-1)n+i,(t-1)n+s} = [\boldsymbol{R}(j - t)]_{is}. \tag{5.14}$$

Accordingly, $\boldsymbol{\Phi}_K$ can be written as a block Toeplitz matrix

$$\begin{pmatrix} R(0) & R(1) & \dots & R(K-1) \\ R(-1) & R(0) & \dots & R(K-2) \\ \vdots & \vdots & \dots & \vdots \\ R(-(K-1)) & R(-(K-2)) & \dots & R(0) \end{pmatrix}. \tag{5.15}$$

Note that $\boldsymbol{R}^{\mathrm{T}}(k) = \boldsymbol{R}(-k)$.

Berger [8, page 111] shows that for an arbitrary correlation matrix $\boldsymbol{\Phi}_K$ the rate distortion function can be defined parametrically in terms of the eigenvalues:

$$D_\theta = K^{-1} \sum_{i=1}^{nK} \min(\theta, \lambda_{i,K}) \tag{5.16}$$

and

$$R_K(D_\theta) = K^{-1} \sum_{i=1}^{nK} \max\left(0, \frac{1}{2}\log\left[\frac{\lambda_{i,K}}{\theta}\right]\right) \tag{5.17}$$

where $\lambda_{i,K}$ is the ith eigenvalue of $\boldsymbol{\Phi}_K$. The sums in (5.16) and (5.17) are divided by K rather than nK because in this monograph the distortion measure is on a per measurement rather than per component basis.

From Cheng [14, Theorem 3.2], for an arbitrary continuous function f operating on a block Toeplitz matrix, we have the asymptotic expression

$$\lim_{K\to\infty} \frac{1}{K} \sum_{i=1}^{nK} f(\lambda_{i,K}) = \frac{1}{2\pi} \int_{-\pi}^{\pi} \mathrm{Trc}\left[f(\boldsymbol{\Phi}(\omega))\right] \mathrm{d}\omega\,, \tag{5.18}$$

provided that f is also valid when λ is replaced by a matrix variable. (Trc[$\cdot$] denotes the Trace of the matrix.) In this equation, $\boldsymbol{\Phi}(\omega)$ is the matrix 'power' spectral function, defined by

$$\boldsymbol{\Phi}(\omega) = \boldsymbol{R}(0) + \sum_{i=1}^{\infty} [\boldsymbol{R}(i)e^{\jmath i\omega} + \boldsymbol{R}(-i)e^{-\jmath i\omega}]. \tag{5.19}$$

It is assumed that $\boldsymbol{\Phi}(\omega)$ is continuous, hermitian and positive definite.

If $\boldsymbol{\Phi}(\omega)$ is hermitian then so is $f(\boldsymbol{\Phi}(\omega))$, so that using standard properties of matrices [37] we have that

$$\mathrm{Trc}\left[f(\boldsymbol{\Phi}(\omega))\right] = \sum_{i=1}^{n} f(\mu_i(\omega)) \tag{5.20}$$

where the $\mu_i(\omega)$ are the eigenvalues of $\boldsymbol{\Phi}(\omega)$.

Now consider the limit as $K \to \infty$ of (5.16) and (5.17). By use of (5.18) and (5.20) we see that

$$D_\theta = \frac{1}{2\pi} \int_{-\pi}^{\pi} \sum_{i=1}^{n} \min(\theta, \mu_i(\omega))\, \mathrm{d}\omega \tag{5.21}$$

and

$$R_\theta = \frac{1}{4\pi}\int_{-\pi}^{\pi}\sum_{i=1}^{n}\max\left(0, \log\left(\frac{\mu_i(\omega)}{\theta}\right)\right) d\omega . \tag{5.22}$$

Consider the case where the position vector $\boldsymbol{X}$ is derived by taking equidistant samples of $\beta(t)$ at intervals δt seconds apart. Following the same process as Berger [8] the rate distortion function for the continuous case can be deduced by considering the limit of the above two equations as δt goes to zero. The result is

$$D_\theta = \frac{1}{2\pi}\int_{-\infty}^{\infty}\sum_{i=1}^{n}\min(\theta, \mu_i(\omega))\, d\omega \tag{5.23}$$

and

$$R_\theta = \frac{1}{4\pi}\int_{-\infty}^{\infty}\sum_{i=1}^{n}\max\left(0, \log\left(\frac{\mu_i(\omega)}{\theta}\right)\right) d\omega \tag{5.24}$$

where the $\mu_i(\omega)$ are the eigenvalues of $\boldsymbol{\Phi}(\omega)$ and $\boldsymbol{\Phi}(\omega)$ is defined as

$$\boldsymbol{\Phi}(\omega) = \int_{-\infty}^{\infty} \boldsymbol{R}(\tau) e^{-\jmath\omega\tau}\, d\tau \tag{5.25}$$

and the variance matrix $\boldsymbol{R}(\tau)$ is defined by

$$[\boldsymbol{R}(\tau)]_{is} = \mathcal{E}\{\boldsymbol{x}_i(t)\boldsymbol{x}_s(t+\tau)\}. \tag{5.26}$$

Example 5.2

Suppose that for an n-dimensional gaussian process the spectral matrix is given by

$$[\boldsymbol{\Phi}]_{is} = \begin{cases} \frac{\pi\sigma^2}{\omega_0} & \text{if } |\omega| \le \omega_0 \text{ and } i = s \\ 0 & \text{otherwise} \end{cases} \tag{5.27}$$

where σ^2 is the variance of the process and ω_0 is the angular cutoff frequency. This would correspond to a vector bandlimited white noise process. The eigenvalues of this matrix will all have the same value: $\frac{\pi\sigma^2}{\omega_0}$.

Substituting these eigenvalues into (5.23) and (5.24) gives

$$D_\theta = \begin{cases} \frac{n\omega_0\theta}{\pi} & \text{if } \theta < \frac{\pi\sigma^2}{\omega_0} \\ n\sigma^2 & \text{otherwise} \end{cases} \tag{5.28}$$

and

$$R_\theta = \begin{cases} \frac{2n\omega_0}{4\pi}\log\left(\frac{\pi\sigma^2}{\omega_0\theta}\right) & \text{if } \theta \le \frac{\pi\sigma^2}{\omega_0} \\ 0 & \text{otherwise} \end{cases} \tag{5.29}$$

so that

$$R_\theta = \begin{cases} nB\log(\frac{n\sigma^2}{D}) & \text{if } 0 \le D \le n\sigma^2 \\ 0 & \text{otherwise} \end{cases} . \tag{5.30}$$

□

5.1.2 Limits on Source Information Rate

The previous section calculated the source information rate for a gaussian source. It is much more difficult to calculate the source information rate for non-gaussian sources, but it is possible to establish limits on the possible values of source information rates. These limits are established in a number of theorems set out below.

Proposition 5.3 *Let $\boldsymbol{X}$ be a time-discrete, stationary, n-dimensional source. Then the mean square error (m.s.e.) rate distortion curve $R(D)$ is bounded from below by $\frac{n}{2}\log(\frac{nQ_1}{D})$ where Q_1 is given by*

$$Q_1 = \frac{e^{\frac{2h}{n}}}{(2\pi e)}, \tag{5.31}$$

and h is the entropy rate of the source i.e.

$$h = \lim_{K\to\infty} \frac{\int \mathrm{d}X\, p_X(X)\log(p_X(X))}{K}. \tag{5.32}$$

Proof

From Berger [8, page 132] we have that the generalised Shannon lower bound can be written as

$$R(D_s) \geq h - h(g_s) \tag{5.33}$$

where

$$D_s = \int \mathrm{d}\boldsymbol{x}\, \rho(\boldsymbol{x}) g_s(\boldsymbol{x}), \tag{5.34}$$

$$h(g_s) = \int \mathrm{d}\boldsymbol{x}\, g_s(\boldsymbol{x}) \log(g_s(\boldsymbol{x})), \tag{5.35}$$

$$g_s(\boldsymbol{x}) = \frac{e^{s\rho(\boldsymbol{x})}}{\int \mathrm{d}z\, e^{s\rho(z)}}, \tag{5.36}$$

and $\boldsymbol{x}$ is an arbitrary representative of the stationary sequence $\boldsymbol{X} = \boldsymbol{x}_1, \ldots, \boldsymbol{x}_K$. The m.s.e. requirement means that

$$\rho(\boldsymbol{x}) = x_1^2 + \ldots + x_n^2. \tag{5.37}$$

From (5.34), (5.36) and (5.37), we have that

$$\int \mathrm{d}\boldsymbol{x}\, (x_1^2 + \ldots + x_n^2) e^{s(x_1^2 + \ldots + x_n^2)} = D, \tag{5.38}$$

so that $s = \frac{n}{2D}$. Substitution into (5.35) gives

$$h(g_s) = \text{entropy rate of } g_s = \frac{n}{2}\log\left(\frac{2\pi e D}{n}\right). \tag{5.39}$$

Accordingly, using (5.31), (5.33) and (5.39) we can write

$$R(D) \geq \frac{n}{2} \log\left(\frac{nQ_1}{D}\right). \tag{5.40}$$

End of Proof

Proposition 5.4

Let X be a zero-mean, time-discrete, n- dimensional vector stationary source with spectral density $\Phi(\omega)$ (see (5.15)). Then $R(D) \leq R_G(D)$, where $R_G(D)$ is the m.s.e. rate distortion function when it is assumed that X_t is gaussian.

Proof

Berger [8, Theorem 4.6.3] has proved this proposition for the scalar case by consideration of the scalar correlation matrix. His argument applies to an arbitrary correlation matrix, so it is simple to generalise to the vector case, demonstrating that

$$R_K(D_\theta) \leq K^{-1} \sum_{i=1}^{nK} \max\left(0, \frac{1}{2} \log\left(\frac{\lambda_{i,K}}{\theta}\right)\right), \tag{5.41}$$

with equality if the source is gaussian (see (5.17)). A consideration of the limit as $K \to \infty$ (see Eqns (5.18) – (5.22)) gives the result that $R(D) \leq R_G(D)$ where $R_G(D)$ is the rate distortion function calculated as if the process X_t were gaussian.

End of Proof

These propositions can be used to provide limits for the important case of bandlimited non- gaussian sources. Consider a stationary vector source $\{\boldsymbol{x}(t)\}$ with matrix spectral density $\Phi(\omega)$, bandlimited to $\omega \leq \omega_0$.

By use of the generalised sampling theorem, it is easy to see that $\boldsymbol{x}(t)$ can be reconstructed from the sequence $\boldsymbol{x}(kT_0)$ (where $k = 0, \pm 1, \ldots$; $T_0 = \frac{\pi}{\omega_0}$) and it is evident that $R(D)$ will be the same for the two cases.

Accordingly, Propositions 5.3 and 5.4 can be readily extended to the continuous case by converting from nats per sample to nats per second. This is done by multiplying by B, the bandwidth in hertz. Note that $B = \frac{\omega_0}{2\pi}$.

We are now in a position to state the following proposition[1].

Proposition 5.5 *Suppose that $(\boldsymbol{X}_t)$ is a continuous stationary vector source with a matrix spectral density that satisfies*

$$[\Phi(\omega)]_{ij} = 0 \text{ if } |\omega| \geq \omega_0;\ i, j = 1, \ldots, n. \tag{5.42}$$

[1] A generalisation of Berger's theorem 4.6.5 [8].

Then the m.s.e. rate distortion function of (X_t) conforms to the following limits:

$$nB\log\left(\frac{nQ_1}{D}\right) \le R(D) \le R_G(D) \tag{5.43}$$

where Q_1 is given by

$$Q_1 = \frac{e^{\frac{h}{nB}}}{2\pi e} \tag{5.44}$$

and h is the entropy rate of the sampled time discrete source.

This proposition follows immediately from Propositions 5.3 and 5.4 by making the appropriate scale conversions.

The work in this chapter could be extended by deriving formulae for non-gaussian sources (e.g. those of Gerrish and Shulteiss [38] and Wyner [130]).

Example 5.6

Consider a one-dimensional positioning system. Suppose that there is one object of interest, with an equation of motion governed by the stochastic differential equation

$$\frac{d^2x(t)}{dt^2} + \beta\frac{dx(t)}{dt} + \omega_0^2 x(t) = n(t) \tag{5.45}$$

where β is proportional to the coefficient of friction, $n(t)$ is a noise source and ω_0 is the angular frequency of oscillation in the absence of friction. This is the equation of motion of a harmonically bound particle (see Papoulis [87, page 524]). Such an equation could be used as a simplified model of a positioning system where the objects have a tendency to return to a home base, can suffer unpredictable accelerations, and are subject to some form of damping.

The transfer function of the linear equation will be given by

$$S_h(\omega) = \frac{1}{(\jmath\omega)^2 + \jmath\omega\beta + \omega_0^2} \tag{5.46}$$

The presence of the restoring force makes this system stationary so that the power spectrum of $x(t)$ is given by

$$S_{xx}(\omega) = \frac{S_{nn}(\omega)}{(\omega^2 - \omega_0^2)^2 + \beta^2\omega^2} \tag{5.47}$$

where S_{nn} is the power spectrum of the noise.

The auto-correlation function $R_{xx}(t)$ can be found by Fourier transforming the spectrum. If $n(t)$ is white gaussian noise with a spectral density of α, the resultant auto-correlation can be divided into three types (oscillatory, over damped and critically damped) depending on the sign of $(\frac{1}{4}\beta^2 - \omega_0^2)$. In all these cases, the probability density function of $x(t)$ is gaussian distributed with variance equal to

$$\sigma_x^2 = R_{xx}(0) = \frac{\alpha}{2\beta\omega_0^2}. \tag{5.48}$$

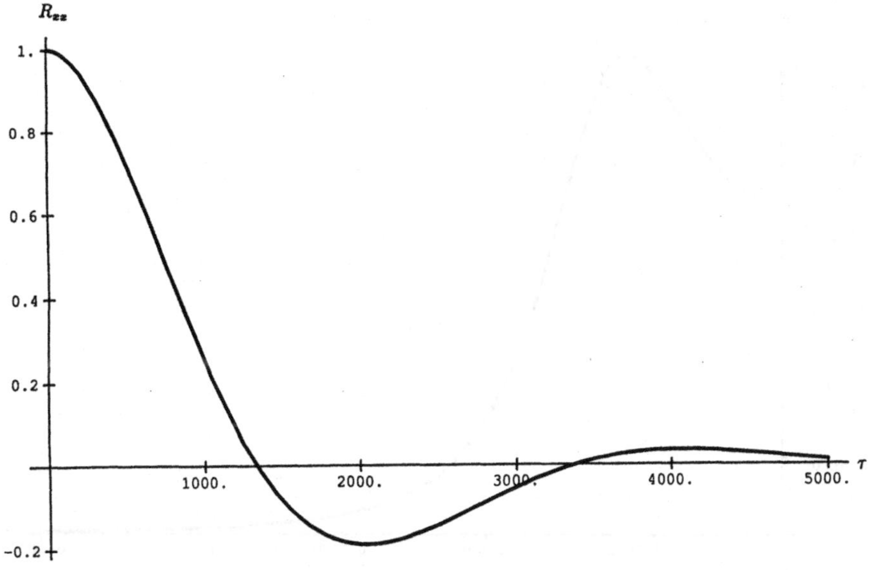

Fig. 5.1. Auto-correlation Function for Harmonically Bound Object

Let us suppose that the standard deviation of the process is known to be 10 kilometres, there is an inherent period of 1 hour, and that the value of α is 1 m^2s^{-3}. Equation (5.48) can be solved to give $\beta \simeq 0.001641$. This means that the motion will be oscillatory with $R_{xx}(t)$ given by

$$R_{xx}(\tau) = \sigma_x^2 e^{-\frac{\beta\tau}{2}} \left(\cos\omega_1\tau + \frac{\beta}{2\omega_1} \sin\omega_1\tau \right) \tag{5.49}$$

where $\omega_1 = \sqrt{\omega_0^2 - \frac{1}{\beta^2}}$.

The function, R_{xx}, normalised by $R_{xx}(0)$ is plotted in Fig. 5.1, and the power spectral density is shown in Fig. 5.2.

The auto-correlation of the velocity will be given by

$$R_{vv}(\tau) = \frac{d^2 R_{xx}(\tau)}{d\tau}, \tag{5.50}$$

so

$$R_{vv}(\tau) = \sigma_v^2 e^{-\frac{\beta\tau}{2}} \left(\cos\omega_1\tau - \frac{\beta}{2\omega_1} \sin\omega_1\tau \right) \tag{5.51}$$

where $\sigma_v = \frac{\alpha}{2\beta}$. With the above values for α and β, $\sigma_v = 17.5$ ms^{-1}.

By using (5.47) to evaluate (5.23) and (5.24) the source information rate can be calculated. A plot of the source information rate as a function of allowable

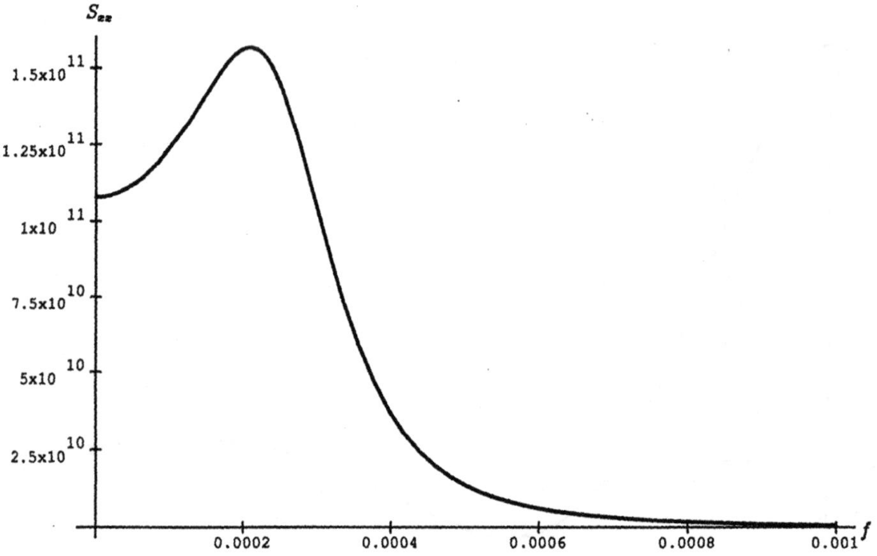

Fig. 5.2. Power Spectral Density for a Harmonically Bound Object

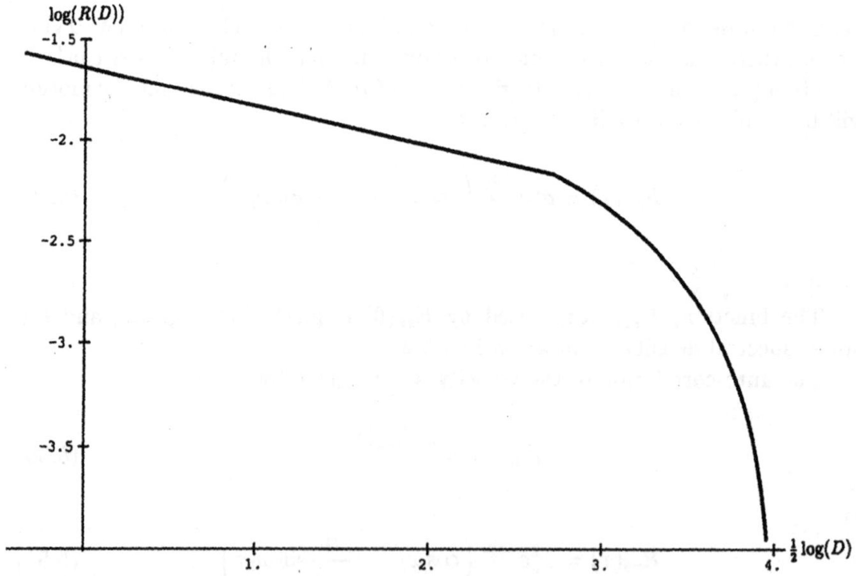

Fig. 5.3. Source Information Rate for Harmonically Bound Object

r.m.s. error is shown in Fig. 5.3. Note that the information rate is very low due to the small effective bandwidth of the process (see Fig. 5.2).

□

A major use of the results in this section could be to compare the source information rate with the predicted performance of a system to estimate how effectively (and safely) the system is being utilised. For example, in an air traffic control radar, this calculation would give an estimate of how much of the capacity of the radar was being used. If performance exceeds capacity, then measurement errors must occur.

The foregoing has shown how it is possible to measure the information rate of stochastic processes. However this analysis cannot be applied to deterministic processes, yet it can occur that the motion of a remote object is best described by deterministic equations. Recently Shaw [104] has shown how it is possible to calculate the information rate of a deterministic system. This has created much interest in the analysis of chaos in dynamical systems [66, 93, 80, 3] and other areas [58, 3].

For deterministic processes it is possible to measure the trajectory of the object with arbitrary accuracy with only a finite channel capacity. The deterministic nature means that by observing the process for a sufficiently long period and combining measurements we can reduce the overall distortion to an arbitrary level. Accordingly we can define an ***absolute source information rate*** $R(0)$: the information rate necessary to measure a deterministic process with arbitrary accuracy. The size of this absolute information rate will be a measure of the degree of chaos inherent in the process.

For a number of cases, it is possible to derive explicit formulations for the source information rate of deterministic processes. For example, if the deterministic process is ergodic, the absolute information rate is given by the Kolmogorov-Sinai entropy [105, 34] (per unit time). Shaw [104] calculates that in the case of a constant velocity object, the information flow (and hence the source information rate) will be

$$R(0) = \frac{1}{t}. \tag{5.52}$$

In many cases, the objects being measured by positioning systems will either be constant velocity, or for significant periods of times be thought of as moving with constant velocity. Accordingly the above equation shows that the source information rate approaches zero if the object is observed for significant periods of time.

Indeed in many cases a tighter bound on the performance of these systems is provided by the source information rate rather than the system capacity. After initial localisation, if the source information rate is smaller than the channel capacity then the limit on performance will be the source information rate. This issue is discussed in more detail in Chap. 6.

Note that these definitions of source information rate (chaotic and stochastic) do not account for the information gained by the initial localisation (acquisition) of the object/image. Accordingly they can be thought of as long term source information rates.

5.2 Wave-Form Coding

Wave-form coding for wave-based positioning systems consists of selection of the appropriate modulation scheme. The waveforms will normally be chosen to optimise aspects of the physical coding (see Sect. 5.3). For example if ranging is being done by time delay measurements, it is normal to choose a modulating waveform whose autocorrelation function has low sidelobes. This minimises the chances of a sidelobe peak being mistaken for the main peak.

It is difficult to understand the implications of choosing various waveforms without first understanding the process of waveform decoding. Accordingly the issue of waveform selection will not be discussed until Chap. 6.

5.3 Physical Coding

A received waveform can be specified in terms of its phase, amplitude and polarization. The process of transmission (reflection in the case of radars and active sonars) and propagation of the wave will alter the values of the phase, amplitude and polarization. The values of these parameters will alter with respect to space, time and frequency. It is the variations of these parameters, as induced by the physical propagation, which is referred to as physical coding. These variations can be ultimately be determined by use of the appropriate wave equations. For example for electromagnetic waves it would be Maxwell's equations or a suitable generalisation [44].

There are a large number of attributes about a moving object that can be transduced by physical coding. These attributes can be classified in the following manner[2]: the receiver will measure the phase, amplitude and polarization of the incoming wave. These parameters can be measured as a function of space, time and frequency. By considering each of the various possibilities the available information can be systematically classified. This is done below:

Absolute Amplitude If the amplitude of the received signal, the amplitude of the emitted signal and the nature of the transmission medium's attenuation are all known, then it is possible to deduce the range of the object. This is normally only practicable for short range systems.

Absolute Phase If the absolute phase of the signal is known, the range can also be deduced. This is possible if the wavelength is much larger than the volume of interest.

[2]This classification is an extension of a method presented by Skolnik [108, page 454].

Polarization If the polarization of the signal is known, and a suitable waveform coding technique is utilised, it is possible to measure the arrival angle and orientation of the remote object. For example the Spasyn system [67] transmits a signal with three identifiable, orthogonal polarizations. By measuring the received polarization of the three transmitted polarizations, it is possible to deduce the position and orientation of the remote transmitter.

Variation of Phase with Position The phase of a wave will vary from point to point. By measuring the nature of this variation while time and frequency are kept constant, it is possible to deduce the arrival angle of the wave. This can be done by using a phased array, some other form of directional antenna, or by sampling the wave-front as in Synthetic Aperture Radar (SAR) [45].

Variation of Phase with Time Another source of phase variation will be due to a relative velocity between the object and the receiver. The Doppler effect will cause a frequency shift, which is simply a linear variation of phase with time. By measuring the frequency shift while keeping position and frequency constant, it is possible to deduce the relative velocity.

Variation of Phase with Frequency The phase shift due to finite propagation can be expressed in terms of the number of wavelengths travelled by the wave. This will differ at different frequencies. By measuring the phase variation as a function of frequency while keeping time and position constant, the range of the object can be deduced. In the case of a non-dispersive medium there will be a linear relationship between the phase and the frequency. The slope of this relationship will be proportional to the distance between the object and receiver. The linear relationship between phase and frequency in the frequency domain implies a time shift in the time domain. Accordingly a receiver that estimates range by measuring time shifts is using the variation of phase with frequency.

Variation of Amplitude with Position To first order, the amplitude of the received signal will vary because of the different aspects that the object can present. Accordingly this variation can be used to deduce information about the shape of the object, provided time and frequency are kept constant. If the shape is known then it may be possible to deduce the orientation. A pure positioning system will only be concerned with the orientation.

Variation of Amplitude with Time Over short time periods, the major contributor to changes in amplitude will be rotation or change of shape of the object, provided position and frequency are kept constant. A pure positioning system can estimate the rotation of the object if the shape is known.

Variation of Amplitude with Frequency For a reflective system (e.g. radar or active sonar), the amplitude of the reflected signal will depend on the frequency. This is because the scattering cross-section depends on the wavelength. For long wavelengths (compared to the object size) the predominant scattering will be Rayleigh, while for short wavelengths simple reflection will be the major contributor to the returned signal. Accordingly, measurements of amplitude as a function of frequency could be used to estimate target size. A pure positioning system would not use this information.

Variation of Polarization with Position The received polarization will vary as a function of position. This variation could be used to measure the position of the object, provided time and position are kept constant. The basic method would be triangulation.

Variation of Polarization with Time Over short time periods, the variation of polarization with time will be due to rotation or change of shape of the object. A pure positioning system can therefore use measurements of polarization as a function of time to deduce information concerning the rotation of a remote object. In this case position and frequency should be kept constant.

Variation of Polarization with Frequency For a reflective system, the received polarization could vary with frequency due to frequency dependent anisotropies. In principal this information could be used to deduce the shape of the object. It is unlikely to be used in a pure positioning system.

Example 5.7

Consider a two-dimensional remote positioning system (see Fig. 5.4), with the remote object located at (x_1, y_1) and a reception station at (x_0, y_0). Consider a transmitter that emits a signal $e^{j2\pi f_0 t}$ which is received by an omni-directional antenna. The phase of the received signal, η, can be represented by

$$\eta = 2\pi\left(\frac{D_r}{c}f_0 + f_0 t\right) \tag{5.53}$$

where c is the wave velocity and D_r is the distance between the transmitter and the reception station. Note that to simplify the following discussion, ambiguities will be neglected.

Now consider the partial derivatives of η with respect to t, f_0 and x_0.

$$\frac{\partial\eta}{\partial t} = 2\pi f_0\left(1 + \frac{\partial D_r}{c\partial t}\right), \tag{5.54}$$

$$\frac{\partial\eta}{\partial f_0} = 2\pi\left(\frac{D_r}{c} + t\right), \tag{5.55}$$

and

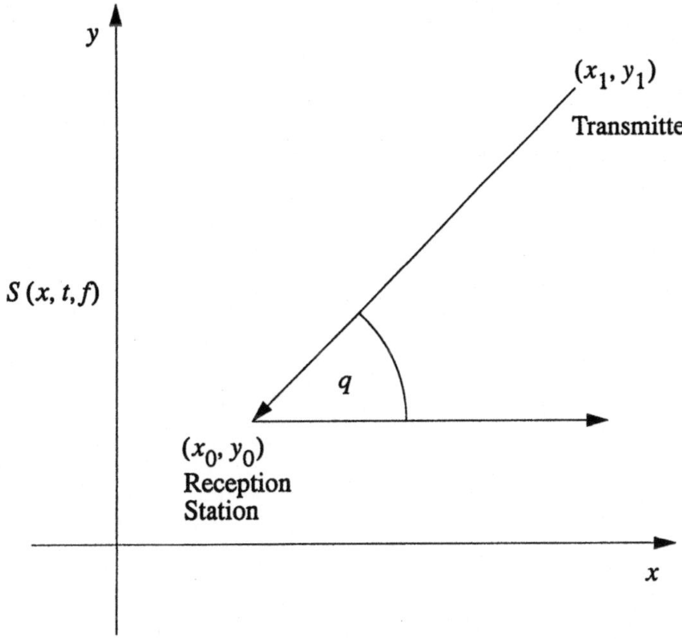

Fig. 5.4. Geometry of Two-Dimensional Remote Positioning System

$$\frac{\partial \eta}{\partial x_0} = 2\pi f_0 \frac{(x_1 - x_0)}{cD_r} \tag{5.56}$$

or

$$\frac{\partial \eta}{\partial x_0} = \frac{2\pi f_0}{c} \sin\theta, \tag{5.57}$$

where θ is the arrival angle (see Fig. 5.4). Now $\frac{\partial D_r}{\partial t}$ is simply the radial velocity which can be estimated using (5.54). If there is a time reference, then the radial distance, D_r, can be deduced using (5.55) while the arrival angle can be estimated using (5.57).

This simple example has demonstrated how the partial phase derivatives can be used to estimate radial velocity, radial distance and arrival angle. □

6. Decoding

There are two aspect to the decoding process: physical decoding and waveform decoding. Physical decoding is the process of transforming the measurements made in the communications frame to the world frame. Provided there is a 1:1 mapping between the two frames this is a very simple process, so physical decoding will not be discussed further.

The process of waveform decoding requires a receiver to estimate various signal parameters. This chapter first derives the optimal receiver for discrete parameters (Sect. 6.1) and then extends this result to continuous parameters (Sect. 6.2). Following this, various ways of characterising waveform decoding are considered including accuracy, noise ambiguity and signal ambiguity (Sect. 6.3 and 6.5). Then a definition of the performance of a particular coding scheme is developed (Sect. 6.6). Finally the research areas of realistic channels (Sect. 6.8.1) and waveform selection (Sect. 6.8.2) are addressed.

6.1 Discrete Optimal Receiver

This section derives the optimal receiver for the reception of discrete parameters. The arguments are generalisations of well known results for one-dimensional channels [64]. In the process it is demonstrated how a data rate and associated error rate can be defined for a positioning system. Those readers familiar with the optimal receiver concept may proceed directly to Sect. 6.1.1.

The following assumptions are made in the ensuing derivation: each measurement is independent, only one measurement is made at a time, and the *a priori* p.d.f. for each object is uniformly distributed (in the communications frame) over the domain of interest.

In the communications frame, the aim is to transmit the position $\boldsymbol{\xi}$ over the channel. This is done by encoding the information as a multi-dimensional signal $u(\boldsymbol{w}, \boldsymbol{\xi})$, a multi-link signal $u_1(w_1, \xi_1), \ldots, u_n(w_n, \xi_n)$, or perhaps some combination. For example, a two-dimensional polar positioning system might encode the position as a function of time and angle, so that the signal could be represented as $u = (\omega_1, \omega_2, \xi_1, \xi_2) = u(\omega_1 - \xi_1, \omega_2 - \xi_2)$, where ω_1 is time, ω_2 is angle, ξ_1 is the actual time delay of the object and ξ_2 is the actual angle of the object. In what follows it will be assumed that the signal is multi-dimensional. It is easy to adapt the results to the multi-link case.

The parameter $\boldsymbol{\xi}$ will be constrained to a volume of interest U_n. Let us divide this volume into M different segments, and label them $1,\ldots,M$. For each of these segments, choose an interior point to the segment, $\boldsymbol{\xi}_i$, where $i = 1,\ldots,M$. The interior point could be the centroid. Assume that the transmitter sends positions which correspond exactly to the ξ_i of one of the boxes (i.e. the possible positions are quantised). The receiver could estimate what value of $\boldsymbol{\xi}$ was transmitted, by guessing which of the M boxes the position lies in. Without loss of generality, assume that the true position is in the first box.

If the transmitted signal $u(\boldsymbol{w},\boldsymbol{\xi}_i)$ is corrupted by white zero-mean gaussian noise $n(\boldsymbol{w})$ then the received signal v is given by

$$v(\boldsymbol{w}) = u(\boldsymbol{w},\boldsymbol{\xi}_1) + n_o(\boldsymbol{w}). \tag{6.1}$$

Here the effects of attenuation on the signal are ignored[1].

Assume that $v(\boldsymbol{w},\boldsymbol{\xi})$,$u(\boldsymbol{w},\boldsymbol{\xi}_i)$ and $n_o(\boldsymbol{w})$ can be represented by as points in some suitable N-dimensional signal space, and denote these points as $\boldsymbol{v}$,$\boldsymbol{u}^i$ and $\boldsymbol{n}_o$ respectively. For example if the signals are a function of one variable, the signal space representation could be simply a vector of time samples: $\boldsymbol{v} = (v(t_1,\xi)\ldots v(t_N,\xi))$. If the signals are a function of two variables (say time and angle), the enumeration would have to cover the area of interest for both the variables.

When placed in this context, the reception problem is to determine to which of the boxes the signal corresponds. One way to decode the signal is to determine which is the most probable message given that $\boldsymbol{v}$ is received. Accordingly the optimal receiver maximises the *a posteriori* probability distribution $P(\boldsymbol{u}^i|\boldsymbol{v})$: the probability that $\boldsymbol{u}^i$ was sent given that $\boldsymbol{v}$ was received. See Chap. 7 for a discussion on other optimality criteria.

Using Bayes rule we can write

$$P(\boldsymbol{u}^i|\boldsymbol{v}) = \frac{P(\boldsymbol{v}|\boldsymbol{u}^i)P(\boldsymbol{u}^i)}{P(\boldsymbol{v})}. \tag{6.2}$$

Now $P(\boldsymbol{v}|\boldsymbol{u}^i)$ is the probability of receiving $\boldsymbol{v}$ when $\boldsymbol{u}^i$ was sent. It can be written as

$$P(\boldsymbol{v}|\boldsymbol{u}^i) = \frac{1}{\sigma_{n0}^M (2\pi)^{\frac{M}{2}}} \exp\left(\frac{-\sum_{r=1}^M |(v_r - u_r^i)|^2}{(2\sigma_{n0}^2)}\right) \tag{6.3}$$

where σ_{n0}^2 is the variance of the noise components. Note that the noise is considered to have a bandwidth much wider than the receiver, but still have finite variance. By considering the limit as the number of subdivisions M goes to infinity, this equation becomes

$$P(\boldsymbol{v}|\boldsymbol{u}^i) = k_a \exp\left(-\frac{1}{N_g} < |(\boldsymbol{v} - \boldsymbol{u}^i)|^2 >\right) \tag{6.4}$$

[1]The effects of uncertainty in the amplitude or frequency of the signal can be included by augmenting w with the additional parameters.

where k_a is a normalisation constant, $< \cdot >$ is the signal space inner product [33], and N_g is equal to σ_n^2 divided by the product of the bandwidths. Suppose the n-dimensional channel has dimensions of view L_i and bandwidths B_i (see (3.24)). Then

$$N_g = \frac{\sigma_n^2}{\prod_{i=1}^{n} B_i}. \tag{6.5}$$

The channels have been numbered so that the nth channel is the time channel (i.e. $B_n = B$ and $L_n = T$).

The inner product is defined as

$$< \boldsymbol{v}\boldsymbol{u}^i >= \int \ldots \int \mathrm{d}\boldsymbol{w}\, v(\boldsymbol{w})u^*(\boldsymbol{w}, \boldsymbol{\xi}_i) \tag{6.6}$$

where the limits of integration will be over the area where the signals are non-zero. For example, if the signal space is functions of time and the received signal is observed for a time T, then

$$< vu^i >=< v(t)u(t, \xi_i) >= \int_0^T \mathrm{d}t\, v(t)u(t, \xi_i). \tag{6.7}$$

Substituting (6.4) into (6.2) gives

$$P(\boldsymbol{u}^i|\boldsymbol{v}) = \frac{k_a P(\boldsymbol{u}^i) \exp\left(-\frac{1}{N_g} < |(\boldsymbol{v} - \boldsymbol{u}^i)|^2 >\right)}{P(\boldsymbol{v})}. \tag{6.8}$$

Now let us evaluate $P(\boldsymbol{v})$, using

$$P(\boldsymbol{v}) = \sum_{j=1}^{M} P(\boldsymbol{u}^j)P(\boldsymbol{v}|\boldsymbol{u}^j). \tag{6.9}$$

Use of (6.4) gives

$$P(\boldsymbol{v}) = k_a \sum_{j=1}^{M} P(\boldsymbol{u}^j) \exp\left(-\frac{1}{N_g} < |\boldsymbol{v} - \boldsymbol{u}^j|^2 >\right). \tag{6.10}$$

Substituting this equation into (6.8) yields

$$P(\boldsymbol{u}^i|\boldsymbol{v}) = \frac{P(\boldsymbol{u}^i) \exp\left(-\frac{1}{N_g} < |\boldsymbol{v} - \boldsymbol{u}^i|^2 >\right)}{\sum_{i=1}^{M} P(\boldsymbol{u}^j) \exp\left(-\frac{1}{N_g} < (\boldsymbol{v} - \boldsymbol{u}^j)^2 >\right)}. \tag{6.11}$$

Note that the denominator does not depend on $\boldsymbol{u}^i$. We have assumed that all the signals are equiprobable, so the maximum value of $P(\boldsymbol{u}^i|\boldsymbol{v})$ will occur when $< |\boldsymbol{v} - \boldsymbol{u}^i|^2 >$ is minimised. This is simply a statement that the best choice is the $\boldsymbol{u}^i$ that is closest (in the euclidean sense) to $\boldsymbol{v}$. Let us expand $< |\boldsymbol{v} - \boldsymbol{u}^i|^2 >$:

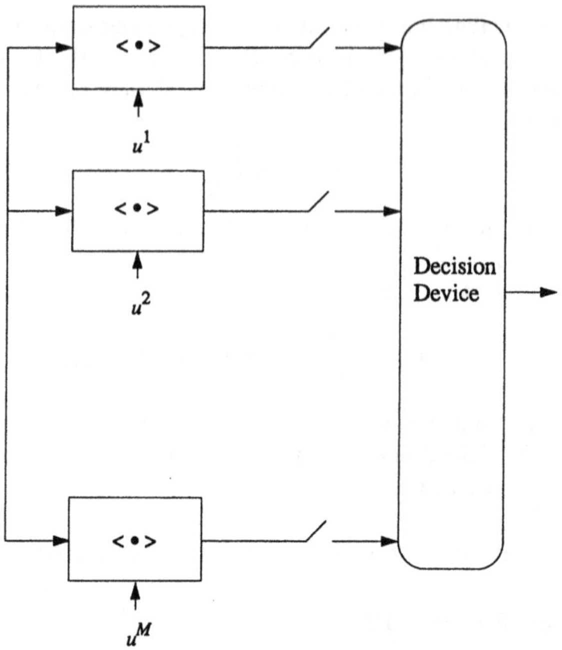

Fig. 6.1. The Ideal Receiver

$$< |\boldsymbol{v} - \boldsymbol{u}^i|^2 >=< \boldsymbol{v}\boldsymbol{v} > + < \boldsymbol{u}^i\boldsymbol{u}^i > -2 < \boldsymbol{v}\boldsymbol{u}^i > . \tag{6.12}$$

The first term on the right is the energy in $\boldsymbol{v}$ and is independent of i, the second term is the energy in $\boldsymbol{u}^i$. Assume $\boldsymbol{u}^i = u(\boldsymbol{w}, \boldsymbol{\xi}_i)$ is given by $u(\boldsymbol{w} - \boldsymbol{\xi}_i)$, so that each of the $\boldsymbol{u}^i$ will have equal energy and $< \boldsymbol{u}^i\boldsymbol{u}^i >$ will be independent of i. Accordingly, to minimise $< (\boldsymbol{v} - \boldsymbol{u}^i)^2 >$ it is necessary only to choose $\boldsymbol{u}^i$ to maximise

$$< \boldsymbol{v}\boldsymbol{u}^i > . \tag{6.13}$$

This implies that the ideal receiver for this sort of signal will be a series of correlators, as shown in Fig. 6.1.

Denote the output of the ith correlator as z_i:

$$z_i = \frac{< \boldsymbol{v}\boldsymbol{u}^i >}{T_g} \tag{6.14}$$

where T_g is the observation volume

$$T_g = \prod_{i=1}^{n} L_i . \tag{6.15}$$

If the channel is a one-dimensional time channel, then $T_g = T$, the observation time.

The ideal receiver will examine each of the z_i and determine which is the largest. In our case the largest should be z_1 and the object will be assumed to be located at $\boldsymbol{\xi}_1$. The z_i are random variables, so it is possible to make an error in this assignment. Call the probability of an error P_E and the probability of being correct P_C.

The probability of being correct is the probability that z_1 is greater than $z_2, \ldots, z_M$ so that

$$P_C = \int_{-\infty}^{\infty} \mathrm{d}z_1 \int_{-\infty}^{z_1} \cdots \int_{-\infty}^{z_1} \mathrm{d}z_2 \ldots \mathrm{d}z_M \, p(z_1, \ldots, z_M), \tag{6.16}$$

where $p(z_1, \ldots, z_M)$ is a gaussian probability density function given by

$$p(\boldsymbol{z}) = \frac{\det(\boldsymbol{A}^{-1})}{(2\pi)^{\frac{M}{2}}} \exp\left(\frac{1}{2}(\boldsymbol{z} - \overline{\boldsymbol{z}})^{\mathrm{T}} \boldsymbol{A}^{-1} (\boldsymbol{z} - \overline{\boldsymbol{z}})\right), \tag{6.17}$$

where $\boldsymbol{z}$ denotes $z_1, \ldots, z_M$, $\overline{\boldsymbol{z}}$ is the expectation of $\boldsymbol{z}$, and $\boldsymbol{A}$ is the covariance matrix:

$$[A]_{ij} = \overline{z_i z_j} - \overline{z_i}\,\overline{z_j}. \tag{6.18}$$

The error probability is given by

$$P_E = 1 - P_C. \tag{6.19}$$

In order to evaluate (6.16), it is necessary to calculate the means and correlation coefficient matrix for $\boldsymbol{z}$. Define

$$\rho_{ij} = \frac{< \boldsymbol{u}^i \boldsymbol{u}^j >}{\sqrt{< \boldsymbol{u}^i \boldsymbol{u}^i >< \boldsymbol{u}^j \boldsymbol{u}^j >}}. \tag{6.20}$$

Now

$$\overline{z_i} = \frac{\overline{< \boldsymbol{v} \boldsymbol{u}^i >}}{T_g} = \frac{1}{T_g} < (u(\mathbf{w}, \boldsymbol{\xi}_1) + \overline{n(\mathbf{w})}) u(\mathbf{w}, \boldsymbol{\xi}_i) > , \tag{6.21}$$

so that noting the noise has a zero mean gives

$$\overline{z_i} = \frac{1}{T_g} < u(\mathbf{w}, \boldsymbol{\xi}_1) u(\mathbf{w}, \boldsymbol{\xi}_i) >= S_p \rho_{1i} \tag{6.22}$$

where S_p is the signal power (note the assumption that each of the signals has the same energy).

The covariance matrix can be evaluated using (6.18) and (6.14)[2].

$$\begin{aligned} [\boldsymbol{A}]_{ij} &= \frac{1}{T_g^2} \int\int \mathrm{d}\mathbf{w}\, \mathrm{d}\omega\, \overline{[u(\mathbf{w}, \boldsymbol{\xi}_1) + n(\mathbf{w})] u(\mathbf{w}, \boldsymbol{\xi}_i) [u(\omega, \boldsymbol{\xi}_1) + n(\omega)] u(\omega, \boldsymbol{\xi}_j)} \\ &\quad - \frac{1}{T_g^2} \int \mathrm{d}\mathbf{w}\, u(\mathbf{w}, \boldsymbol{\xi}_1) u(\mathbf{w}, \boldsymbol{\xi}_j) \int \mathrm{d}\omega\, u(\omega, \boldsymbol{\xi}_1) u(\omega, \boldsymbol{\xi}_j), \end{aligned} \tag{6.23}$$

[2] This is a generalisation of an argument by Viterbi [122].

where the limits of integration are chosen to cover the observation area. This becomes

$$[\boldsymbol{A}]_{ij} = \frac{1}{T_g^2} \int\int \mathrm{d}\mathbf{w}\ \mathrm{d}\boldsymbol{\omega}\ \boldsymbol{u}(\mathbf{w}, \boldsymbol{\xi}_i)\boldsymbol{u}(\boldsymbol{\omega}, \boldsymbol{\xi}_j)\overline{n(\mathbf{w})n(\boldsymbol{\omega})}. \tag{6.24}$$

Using the fact that for white noise

$$\overline{n(\mathbf{w})n(\boldsymbol{\omega})} = \frac{N_g}{2}\delta(\mathbf{w} - \boldsymbol{\omega}), \tag{6.25}$$

(6.23) reduces to

$$[\boldsymbol{A}]_{ij} = \frac{N_g}{2T_g^2} \int \mathrm{d}\mathbf{w}\ u(\mathbf{w}, \boldsymbol{\xi}_i)u(\mathbf{w}, \boldsymbol{\xi}_\mathbf{j}) = \frac{N_g S_p}{2T_g}\rho_{ij}. \tag{6.26}$$

The variance of the $\boldsymbol{z}_i$ will be simply

$$\sigma_{z_i}^2 = \overline{\boldsymbol{z}_i^2} = \frac{N_g S_p}{2T}. \tag{6.27}$$

By means of the substitution

$$v_i = \frac{z_i - \overline{z_i}}{\sigma_{z_i}^2}, \tag{6.28}$$

and, by use of (6.26) and (6.27), (6.16) becomes

$$P_C = \int_{\infty}^{\infty} \mathrm{d}v_1 \int_{-\infty}^{v_1+\sqrt{\frac{2S_pT_g}{N_g}}(1-\rho_1 2)} \cdots \int_{-\infty}^{v_1+\sqrt{\frac{2S_pT_g}{N_g}}(1-\rho_1 M)} \mathrm{d}v_2 \ldots \mathrm{d}v_M \cdot \frac{\det(\boldsymbol{R}^{-1}) exp(-\frac{1}{2}\boldsymbol{v}^{\mathrm{T}}\boldsymbol{R}^{-1}\boldsymbol{v})}{(2\pi)^{\frac{M}{2}}} \tag{6.29}$$

where $[\boldsymbol{R}]_{ij} = \rho_{ij}$.

6.1.1 Bit Rate

We are now in a position to define the bit rate and the error probability of a positioning system. The number of messages being sent is M. We have assumed that the objects are positioned uniformly over the region of interest, so each message is equiprobable. Accordingly the number of bits being transmitted is $\log_2 M$, and this information is sent in a time T. Hence we can state that the bit rate b_r is given by

$$b_r = \frac{\log_2 M}{T}. \tag{6.30}$$

The probability of making an error is given by (6.19) and (6.29).

In many practical situations the M messages are almost orthogonal, so that $\boldsymbol{R}$ is approximately equal to the identity matrix. Then (6.29) can be simplified further to give [5]

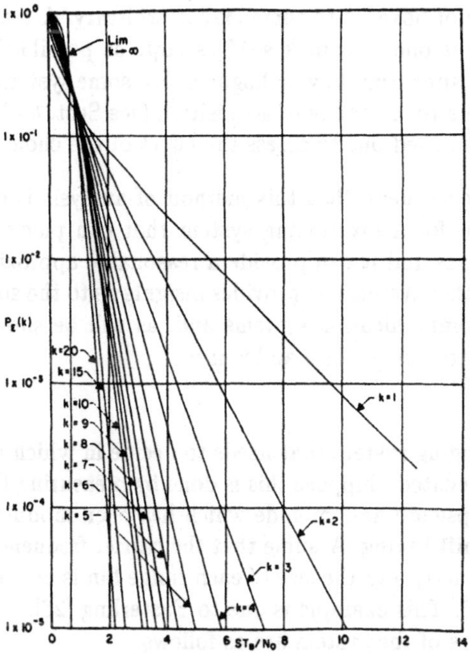

Fig. 6.2. Performance plot for Orthogonal Codes - from 'Digital Communications with Space Applications', editor S. W. Golomb, Prentice Hall, 1964

$$P_E = 1 - P_C = 1 - \int_{-\infty}^{\infty} \frac{exp(\frac{1}{2} - v_1^2)}{\sqrt{2\pi}} \left[\frac{1 + \mathrm{erf}\left(\frac{1}{\sqrt{2}}\left(v_1 + \sqrt{\frac{2S_pT_g}{N_g}}\right)\right)}{2} \right]^{M-1} dv_1 \, . \tag{6.31}$$

Viterbi [122] has plotted the error probability as a function of $\frac{S_pT}{N_0 \log_2 M}$ for different values of $\log_2 M$. His graph has been reproduced in Fig. 6.2 with $k = \log_2 M$ and $T_B = \frac{T}{k}$.

The above demonstrates that for this type of positioning system we have provided a method of defining the bit rate and error probability. Using a suitable model for the error mechanism, it should be possible to combine the bit rate and the error probability into an estimate of the system information rate (see Sect. 6.6).

However it is important to understand the nature of the errors arising from treating a continuous system in a discrete manner:

- There will be a misalignment, so that actual position of the object may lie in the first box but will not be exactly equal to $\boldsymbol{\xi}_1$. Call this quantisation error. This means that the incoming signal will not correspond exactly to one of the M replicas (i.e. the $\boldsymbol{u}^i$), so that error probability calculation will be slightly incorrect, due to the correlation function being non-zero in more than one box.

- The number of boxes is to some extent arbitrary. A reasonable choice for the box size is one that makes M as large as possible while keeping the signal set u^i approximately orthogonal. For some systems, the choice of M is obvious due to the mode of acquisition (see Sect. 6.6). Further research needs to be carried out to assess the effect of the choice of M on P_E.

These two effects mean that this method of analysis is only approximate. It would be useful for a positioning system that can place its measurements into a series of boxes and it can provide a reasonable approximation to system performance for other systems. It provides insights as to the similarities between positioning and communication systems and, as will be shown in Sect. 6.6, it can be used to characterise noise ambiguity.

Example 6.1

Consider a ranging system that needs to decide in which range bin a spacecraft is currently located. Suppose this is done by measuring the time delay in a maximum length pseudorandom code which has been modulated onto the carrier using phase shift keying. Assume that the carrier frequency of the incoming signal is known exactly and the size of each range bin is equal to the size of the chips in the code [3]. This example is due to Easterling [26].

The parameters of the system are as follows:

- receiver noise temperature is 300° K,
- received signal power is -130 dBm,
- maximum Length code with a length of 511 chips,
- the chip period is 1 microsecond, and
- the word error rate (P_E) is 0.0001.

In this case the region of interest is the repetition interval of the code. Obviously the size of the box should be chosen equal to a range bin which means that the signals are nearly orthogonal. This is because a maximal length sequence of length M has an autocorrelation function of 1 when aligned, and $-\frac{1}{M}$ elsewhere.

Now let us evaluate the bit rate of the system, using (6.31) and Fig. 6.2. The value of M is 511 so that $\log_2 M \simeq 9$. Reading off the graph for $k = 9$ and $P_E = 0.0001$ gives a value of 2.8 for $\frac{S_p T_B}{N_0}$. Substituting our numerical value of S_p and using the relationship that N_0 equals Boltzmanns constant times the absolute temperature we arrive at a value of T_B equal to 0.000115. But T_B is simply the time to transmit one bit of information, so the bit rate is the inverse of T_B. Accordingly the system bit rate is 8690 bit/s with a word error rate of 0.0001. □

[3] In order to avoid confusion, the individual bits that make up the code will be referred to as chips, see Holmes [46].

The above analysis could also be used in a more traditional manner to calculate the number of position measurements per second that could be made to a given accuracy. At this point it is worth reiterating two of the reasons for the emphasis on bit-per-second measurements of performance in this monograph. First, if one is comparing two different positioning systems with two different measurement rates and accuracies, calculating the performance in terms of bits per second allows a comparison. Second, if the positioning system performance is specified in bits per second it is possible to make a direct comparison with a communications system.

6.2 Continuous Optimal Receiver

In this section we derive the optimum receiver for continuous parameters. The results from this section will be used in later sections. The reader may omit this section if they are familiar with the continuous optimal receiver. The approach taken here in deriving the optimum receiver is to use the result from Sect. 6.1 [4], where the optimum receiver was derived by dividing the volume of interest into M compartments, labelled $1, \dots, M$. An interior point for each compartment was chosen to give a set of points $\boldsymbol{\xi}_1, \dots, \boldsymbol{\xi}_M$. If the parameter to be transmitted was closest to $\boldsymbol{\xi}_i$, then the signal $\boldsymbol{u}^i$ corresponding to $\boldsymbol{\xi}_i$ would be sent. Our approach here will be to consider the limit as M becomes large.

Suppose that the signal $\boldsymbol{v}$ is received. Given $\boldsymbol{v}$, the probability that $\boldsymbol{\xi}$ lies in the range, $\boldsymbol{\xi}' < \boldsymbol{\xi} < \boldsymbol{\xi}''$ (i.e. $\xi_r' < \xi_r < \xi_r''$; for $r = 1, \dots, n$) is equal to

$$P(\boldsymbol{\xi}' < \boldsymbol{\xi} < \boldsymbol{\xi}''|\boldsymbol{v}) = \sum_{q \in \mathcal{G}} P(\boldsymbol{u}^q|\boldsymbol{v}) \tag{6.32}$$

where $\mathcal{G}$ is the set of compartments which cover the region $\boldsymbol{\xi}' < \boldsymbol{\xi} < \boldsymbol{\xi}''$.

Substituting (6.11) into the above equation gives

$$P(\boldsymbol{\xi}' < \boldsymbol{\xi} < \boldsymbol{\xi}''|\boldsymbol{v}) = \frac{\sum_{q \in \mathcal{G}} P(\boldsymbol{u}^q) \exp\left(-\frac{1}{N_g} < |(\boldsymbol{v} - \boldsymbol{u}^q)|^2 >\right)}{\sum_{i=1}^{M} P(\boldsymbol{u}^j) \exp\left(-\frac{1}{N_g} < |(\boldsymbol{v} - \boldsymbol{u}^j)|^2 >\right)}. \tag{6.33}$$

Let $\boldsymbol{\xi}'' = \boldsymbol{\xi}' + \boldsymbol{\delta\xi}$, multiply top and bottom of (6.33) by $\boldsymbol{\delta\xi}$ and consider the limit as $\boldsymbol{\delta\xi} \to 0$. The numerator can be simply evaluated at $\boldsymbol{\xi}$, giving

$$p_{\xi}(\boldsymbol{\xi}) \exp\left(-\frac{1}{N_g} < |v(\boldsymbol{w}) - u(\boldsymbol{w}, \boldsymbol{\xi})|^2 >\right) \mathrm{d}\boldsymbol{\xi}\,, \tag{6.34}$$

while the sum in the denominator becomes an integral:

$$\int \mathrm{d}\boldsymbol{\mu}\, p_{\mu}(\boldsymbol{\mu}) \exp\left(-\frac{1}{N_g} < (v(\boldsymbol{w}, \boldsymbol{\mu}) - u(\boldsymbol{w}, \boldsymbol{\mu}))^2 >\right) \tag{6.35}$$

[4] The methodology used here is based on the approach used by Kotel'nikov [64].

where the limits of integration are over the volume of interest. Accordingly

$$P(\boldsymbol{\xi}' < \boldsymbol{\xi} < \boldsymbol{\xi}''|\boldsymbol{v}) = \frac{p_\xi(\boldsymbol{\xi})\exp(-\frac{1}{N_g} < |v(\boldsymbol{w}) - u(\boldsymbol{w},\boldsymbol{\xi})|^2 >)\,\mathrm{d}\boldsymbol{\xi}}{\int \mathrm{d}\boldsymbol{\mu}\, p_\mu(\boldsymbol{\mu})\exp(-\frac{1}{N_g} < (v(\boldsymbol{w}) - u(\boldsymbol{w},\boldsymbol{\mu}))^2 >)} \tag{6.36}$$

or

$$p_{\xi|v}(\boldsymbol{\xi}|\boldsymbol{v}) = \frac{p_\xi(\boldsymbol{\xi})\exp\left(-\frac{1}{N_g} < |v(\boldsymbol{w}) - u(\boldsymbol{w},\boldsymbol{\xi})|^2 >\right)}{\int \mathrm{d}\boldsymbol{\mu}\, p_\mu(\boldsymbol{\mu})\exp\left(-\frac{1}{N_g} < (v(\boldsymbol{w}) - u(\boldsymbol{w},\boldsymbol{\mu}))^2 >\right)}. \tag{6.37}$$

The optimal choice for $\boldsymbol{\xi}$ is the one that maximises $p_{\xi|v}(\boldsymbol{\xi}|\boldsymbol{v})$.

If $p_\xi(\boldsymbol{\xi})$ is a uniform distribution and $u(\boldsymbol{w},\boldsymbol{\xi}) = u(\boldsymbol{w} - \boldsymbol{\xi})$ then by the same reasoning as used to arrive at (6.13), we can see that $\boldsymbol{\xi}$ should be chosen to maximise

$$< v(\boldsymbol{w})u(\boldsymbol{w} - \boldsymbol{\xi}) > . \tag{6.38}$$

Note that even if $p(\boldsymbol{\xi})$ is not uniform, the only signal processing operation that the receiver needs to perform on the received signal is cross-correlation, though in the non-uniform case the optimal value may not correspond to the largest correlation.

Example 6.2

Consider a ranging system which measures the time delay of a pulse with a uniform positional p.d.f. The received waveform will be $v(t)$ and the internal replica will be $u(t-\tau)$. Suppose the signal is distorted by additive, white gaussian noise with spectral density N_0. In this case $N_g = N_0$ and the *a posteriori* p.d.f. becomes (see (6.37))

$$p(\tau|v) = K_v p(\tau)\exp\left(\frac{2}{N_0} < v(t)u(t-\tau) >\right) \tag{6.39}$$

where K_v is a normalisation constant that does not depend on τ. If the signal is much greater than the noise then the maximum value the exponent can assume is

$$R = \frac{2 < u(t)u(t) >}{N_0}. \tag{6.40}$$

Note that $< u(t)u(t) >$ is the signal energy.

Following Woodward [129], we can plot the a posteriori p.d.fs for different values of R. In the absence of any information this must correspond to the uniform positional p.d.f. When $R = \frac{1}{4}$ (Fig. 6.3A) the *a posteriori* p.d.f. is characterised by gently rolling fluctuations, indicating a large ambiguity due to the presence of the noise. At $R = 1$ (Fig. 6.3B) distribution has started to 'curdle' into a few peaks, and so that the p.d.f. is not as well connected. When $R = 4$ (Fig. 6.3C), there are two disconnected peaks. One of the peaks is at the correct position, but the other is at the false position, so there is still appreciable noise

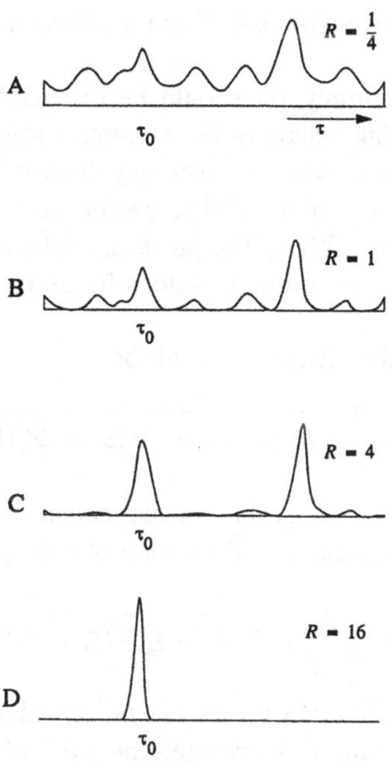

Fig. 6.3. Example *a posteriori* p.d.fs for different values of R

ambiguity. When $R = 16$ (Fig. 6.3D) there is only one narrow peak, indicating an unambiguous well defined position measurement. The width of this peak is an indication of the *accuracy* of the system. The true time delay is marked on the figure as τ_0.

□

The above example demonstrates that two concepts are important when assessing the efficacy of the decoding process, accuracy and noise ambiguity. Accuracy will be discussed in more detail in Sect. 6.3 while noise ambiguity will be discussed in Sect. 6.4.

Of course, we have only derived the optimal receiver for gaussian noise. There is interest in receivers for non-gaussian noise [112, 84, 15, 71, 57, 13], adaptive receivers [85, 121, 119, 120, 11, 43, 111] and receivers implementing other sorts of operations [62, 12, 89].

6.3 Accuracy

An expression for the accuracy of a simple radar system was first derived by Woodward [128]. Slepian [110] also derived the accuracy using the Cramer-Rao bound. In this monograph, the system accuracy will be derived by a generalisation of the work of Kotel'nikov [64]. Complications due to dispersive effects etc. will be ignored.

The actual system accuracy A_w should be calculated in the world frame. However waveform decoding occurs in the communications frame. Accordingly we will have recourse to work with the accuracy defined in the communications frame. Denote this accuracy as A_c. Unlike performance, accuracy will not be invariant, but it is often possible, if the position errors are small, to transform the accuracy measured in the communications frame to an accuracy measured in the world frame.

Based on (2.13), the definition of A_c will be

$$A_c = \frac{1}{K}\sqrt{\mathcal{E}\{(\boldsymbol{\Xi}_i - \widehat{\boldsymbol{\Xi}}_i)^{\mathrm{T}}(\boldsymbol{\Xi}_i - \widehat{\boldsymbol{\Xi}}_i)\}}. \tag{6.41}$$

Assuming the errors from individual measurements will be independent, A_c can be calculated as the square root of the sum of the squares of the individual r.m.s. errors:

$$A_c = \frac{1}{K}\sum_{i=1}^{K}\sqrt{\mathcal{E}\{(\boldsymbol{\xi}_i - \hat{\boldsymbol{\xi}}_i)^{\mathrm{T}}(\boldsymbol{\xi}_i - \hat{\boldsymbol{\xi}}_i)\}}, \tag{6.42}$$

so in the following only individual errors will be considered.

From the previous section we know that the p.d.f. of $\boldsymbol{\xi}$ conditional on $v(\boldsymbol{w})$ is given by

$$p_{\xi|v}(\boldsymbol{\xi}|\boldsymbol{v}) = \frac{p_\xi(\boldsymbol{\xi})\exp\left(-\frac{1}{N_g} < |v(\boldsymbol{w}) - u(\boldsymbol{w},\boldsymbol{\xi})|^2 >\right)}{\int \mathrm{d}\boldsymbol{\mu}\, p(\boldsymbol{\mu})\exp\left(-\frac{1}{N_g} < (v(\boldsymbol{w}) - u(\boldsymbol{w},\boldsymbol{\mu}))^2 >\right)}. \tag{6.43}$$

Let us suppose the $p_\xi(\boldsymbol{\xi})$ is uniformly distributed [5]. We can then write the above equation in the form

$$p_{\xi|v}(\boldsymbol{\xi}|\boldsymbol{v}) = K_v \exp\left(-\frac{1}{N_g} < |v(\boldsymbol{w}) - u(\boldsymbol{w},\boldsymbol{\xi})|^2 >\right) \tag{6.44}$$

where K_v is a constant that does not depend on $\boldsymbol{\xi}$. Denote the value of $\boldsymbol{\xi}$ that maximises this p.d.f. as $\widehat{\boldsymbol{\xi}}$. Note that if this is the method used for estimating $\boldsymbol{\xi}$, then $\widehat{\boldsymbol{\xi}} = \boldsymbol{\phi}$. As discussed earlier, this will occur when $< |v(\boldsymbol{w}) - u(\boldsymbol{w},\boldsymbol{\xi})|^2 >$ is minimum. Hence at $\boldsymbol{\xi} = \widehat{\boldsymbol{\xi}}$, the derivative of this exponent should be zero, so

$$\frac{\partial}{\partial \xi_i} < |v(\boldsymbol{w}) - u(\boldsymbol{w},\boldsymbol{\xi})|^2 > |_{\xi=\widehat{\xi}} = 0;\ i = 1,\ldots,n \tag{6.45}$$

or

$$-2 < u_i'(\boldsymbol{w},\widehat{\boldsymbol{\xi}})[v(\boldsymbol{w}) - u(\boldsymbol{w},\boldsymbol{\xi})] >= 0;\ i = 1,\ldots,n \tag{6.46}$$

where

$$u_i'(\boldsymbol{w},\widehat{\boldsymbol{\xi}}) = \frac{\partial}{\partial \xi_i} u(\boldsymbol{w},\widehat{\boldsymbol{\xi}})|_{\xi=\widehat{\xi}}. \tag{6.47}$$

[5] The argument can be extended to the case where it is not [64].

This equation provides a method of solving for $\hat{\xi}$, and as well this result will be needed in our development.

The accuracy of a positioning system, in the case of high signal-to-noise ratios, will be determined by the spread of $p_{\xi|v}(\xi)$ near $\hat{\xi}$. Taylor-expanding $u(w,\xi)$ in this region:

$$u(\boldsymbol{w},\boldsymbol{\xi}) = u(\boldsymbol{w},\hat{\boldsymbol{\xi}}) - \boldsymbol{u}'^{\mathrm{T}}(\mathbf{w},\hat{\boldsymbol{\xi}})(\boldsymbol{\xi}-\hat{\boldsymbol{\xi}}) + \text{Higher order terms} \tag{6.48}$$

where $\boldsymbol{u}'^{\mathrm{T}} = (u_1', \ldots, u_n')$. Substituting into (6.44) gives

$$\begin{aligned} p_{\xi|v}(\boldsymbol{\xi}|\boldsymbol{v}) &= K_v \exp\Big(-\frac{1}{N_g} < (\, v(\boldsymbol{w}) - u(\boldsymbol{w},\boldsymbol{\xi}) - \boldsymbol{u}'^{\mathrm{T}}(\mathbf{w},\hat{\boldsymbol{\xi}})(\boldsymbol{\xi}-\hat{\boldsymbol{\xi}})\,) \\ &\quad \cdot (\, v(\boldsymbol{w}) - u(\boldsymbol{w},\boldsymbol{\xi}) - \boldsymbol{u}'^{\mathrm{T}}(\mathbf{w},\hat{\boldsymbol{\xi}})(\boldsymbol{\xi}-\hat{\boldsymbol{\xi}})\,) > \Big)\,. \end{aligned} \tag{6.49}$$

Using (6.46) yields

$$p_{\xi|v}(\boldsymbol{\xi}|\boldsymbol{v}) = K_v \exp\left(-\frac{1}{N_g} < (v(\boldsymbol{w}) - u(\boldsymbol{w},\hat{\boldsymbol{\xi}}))^2 > - < |\boldsymbol{u}'^{\mathrm{T}}(\mathbf{w},\hat{\boldsymbol{\xi}})(\boldsymbol{\xi}-\hat{\boldsymbol{\xi}})|^2 > \right) \tag{6.50}$$

or

$$p_{\xi|v}(\boldsymbol{\xi}|\boldsymbol{v}) = K_n \exp\left(-\frac{1}{N_g} < (\boldsymbol{u}'^{\mathrm{T}}(\mathbf{w},\hat{\boldsymbol{\xi}})(\boldsymbol{\xi}-\hat{\boldsymbol{\xi}}))^2 > \right) \tag{6.51}$$

where K_n does not depend on $\boldsymbol{\xi}$.

Using (6.46), this can be written in the form

$$p_{\xi|v}(\boldsymbol{\xi}|\boldsymbol{v}) = K_n \exp\left(-\frac{1}{N_g}(\boldsymbol{\xi}-\hat{\boldsymbol{\xi}})^{\mathrm{T}}\boldsymbol{C}'(\boldsymbol{\xi}-\hat{\boldsymbol{\xi}})\right) \tag{6.52}$$

where K_n is a constant and

$$[C']_{ij} = < u_i'(\boldsymbol{w},\hat{\boldsymbol{\xi}})u_j'(\boldsymbol{w},\hat{\boldsymbol{\xi}}) >;\ i,j = 1,\ldots,n. \tag{6.53}$$

Alternatively

$$p_{\xi|v}(\boldsymbol{\xi}|\boldsymbol{v}) = K_n \exp\left(-\frac{E}{N_g}(\boldsymbol{\xi}-\hat{\boldsymbol{\xi}})^{\mathrm{T}}\boldsymbol{C}(\boldsymbol{\xi}-\hat{\boldsymbol{\xi}})\right) \tag{6.54}$$

where

$$\boldsymbol{C} = \frac{\boldsymbol{C}'}{E} \tag{6.55}$$

and E is the total energy in the signal:

$$E = < u(\boldsymbol{w},\boldsymbol{\xi})^2 > . \tag{6.56}$$

Apart from the constant K_n, this is a gaussian p.d.f. with covariance matrix $\frac{1}{2E}(N_g C^{-1})$ and mean $\hat{\boldsymbol{\xi}}$. Accordingly $\boldsymbol{\xi}$ will be gaussian-like sufficiently close to $\hat{\boldsymbol{\xi}}$. If the signal-to-noise ratio is high enough then the p.d.f. will be very small unless close to $\hat{\boldsymbol{\xi}}$, in which case the distribution will be gaussian everywhere, so K_n can be calculated as a normalisation constant. Note that if $\hat{\boldsymbol{\xi}}$ is near the border of its possible range, then a further analysis needs to be carried out.

When the signal-to-noise ratio is high, the accuracy in the communications frame can be calculated from the variance of the gaussian distribution described in (6.54), giving

$$A_c = \sqrt{\mathcal{E}\{(\boldsymbol{\xi} - \hat{\boldsymbol{\xi}})^{\mathrm{T}}(\boldsymbol{\xi} - \hat{\boldsymbol{\xi}})\}} = \sqrt{\frac{N_g \mathrm{Trc}\left[\boldsymbol{C}^{-1}\right]}{2E}}. \tag{6.57}$$

This will not be the case when the signal to noise ratio is small as noise ambiguity will cause other parts of the *a posteriori* p.d.f. to contribute to the error in A_c.

Normally we will be more interested in the accuracy in the world frame. If the errors are small we can write that $\hat{\boldsymbol{\xi}} - \boldsymbol{\xi} \approx \boldsymbol{J}_m(\boldsymbol{x})(\hat{\boldsymbol{x}} - \boldsymbol{x})$ (see Sect. 4.4). Accordingly, if we transform (6.54) to the world frame the exponent will become $-\frac{E}{N_g}(\boldsymbol{x} - \hat{\boldsymbol{x}})^{\mathrm{T}}\boldsymbol{J}_m(\boldsymbol{x})^{\mathrm{T}}\boldsymbol{C}\boldsymbol{J}_m(\boldsymbol{x})(\boldsymbol{x} - \hat{\boldsymbol{x}})$. Hence

$$A_w = \sqrt{\frac{N_g \mathrm{Trc}\left[\boldsymbol{J}_m^{\mathrm{T}}(\boldsymbol{x})\boldsymbol{C}\boldsymbol{J}_m(\boldsymbol{x})\right]^{-1}}{2E}}. \tag{6.58}$$

In this equation, $\boldsymbol{C}$ is defined by (6.55) and (6.53). We can find a more tractable expression for $\boldsymbol{C}$ in the case when $u(\boldsymbol{w}, \boldsymbol{\xi}) = u(\boldsymbol{w} - \boldsymbol{\xi})$, by defining the Fourier transform pair of a function $\boldsymbol{f}(\boldsymbol{r})$ as follows:

$$U(\boldsymbol{r}) = \boldsymbol{\mathcal{F}}(u(\boldsymbol{w})) =< u(\boldsymbol{w}) \exp(-\jmath 2\pi \boldsymbol{w}^{\mathrm{T}}\boldsymbol{r}) > . \tag{6.59}$$

Use of the power theorem [10] gives

$$[C]_{ij} = \frac{< \mathcal{F}(u_i'(\boldsymbol{w}, \hat{\boldsymbol{\xi}}))\mathcal{F}(u_j'(\boldsymbol{w}, \hat{\boldsymbol{\xi}})) >}{E}; \; i, j = 1, \ldots, n. \tag{6.60}$$

and application of the differentiation theorem [10] yields

$$[C]_{ij} = \frac{< 4\pi^2 r_i r_j |U(\boldsymbol{r})|^2 >}{E}; \; i, j = 1, \ldots, n. \tag{6.61}$$

The quantity, $\mathrm{Trc}\{\boldsymbol{J}_m^{\mathrm{T}}(\boldsymbol{x})\boldsymbol{C}\boldsymbol{J}_m(\boldsymbol{x})\}^{-1}$ (see (6.58)) can be used to define the Geometrical Dilution of Precision (GDOP), which plays an important role in GPS systems [83]. If all the ranging errors are equal, then GDOP is defined as $\mathrm{Trc}\{\boldsymbol{J}_m^{\mathrm{T}}(\boldsymbol{x})\boldsymbol{J}_m(\boldsymbol{x})\}^{-1}$ [61]. Otherwise, GDOP is defined by normalising $\mathrm{Trc}\{\boldsymbol{J}_m^{\mathrm{T}}(\boldsymbol{x})\boldsymbol{C}\boldsymbol{J}_m(\boldsymbol{x})\}^{-1}$ with respect to the average ranging error [72]. GDOP is used in GPS, and other hyperbolic systems to judge the effect of the geometry on the accuracy. Unlike (6.58), it cannot be directly applied to systems that do not have uniform co-ordinates in the communications frame (i.e. a two-dimensional hyperbolic system has units of distance-distance whereas a polar system has units of distance-radians).

Example 6.3

Consider the polar positioning system described in Example 3.3. Denote the frequency response of the system $F(f)$ and the $\frac{x}{\lambda}$ response $M(\frac{x}{\lambda})$ and assume both these are symmetric about zero. The two will be separable so

$$|U(\boldsymbol{r})|^2 = \left|U\left(f, \frac{x}{\lambda}\right)\right|^2 = |F(f)|^2 \left|M\left(\frac{x}{\lambda}\right)\right|^2 . \tag{6.62}$$

Let us define

$$\beta_1^2 \equiv \frac{\int_{-\infty}^{\infty} \mathrm{d}f\, (2\pi f)^2 |F(f)|^2}{\int_{-\infty}^{\infty} \mathrm{d}f\, |F(f)|^2} \tag{6.63}$$

and

$$\beta_2^2 \equiv \frac{\int_{-\infty}^{\infty} \mathrm{d}\frac{x}{\lambda}\, (2\pi(\frac{x}{\lambda}))^2 |M(\frac{x}{\lambda})|^2}{\int_{-\infty}^{\infty} \mathrm{d}\frac{x}{\lambda}\, |M(\frac{x}{\lambda})|^2} . \tag{6.64}$$

The β_i are often referred to as the r.m.s. bandwidths. Using (6.61), (6.63) and (6.64) we have

$$\boldsymbol{C} = \begin{pmatrix} \beta_1^2 & 0 \\ 0 & \beta_2^2 \end{pmatrix} . \tag{6.65}$$

Now $\boldsymbol{J}_m(\boldsymbol{x})$ will be given by

$$\boldsymbol{J}_m(\boldsymbol{x}) = \begin{pmatrix} \frac{x_1}{rc} & \frac{x_2}{rc} \\ -\frac{x_2}{r^2} & \frac{x_1}{r^2} \end{pmatrix} , \tag{6.66}$$

where $r = \sqrt{x_1^2 + x_2^2}$ and c is the velocity of propagation. Accordingly we have that

$$\mathrm{Trc}\{\boldsymbol{J}_m^{\mathrm{T}}(\boldsymbol{x}) \boldsymbol{C} \boldsymbol{J}_m(\boldsymbol{x})\} = \frac{r^2}{\beta_2^2} + \frac{c^2}{\beta_1^2} , \tag{6.67}$$

so using (6.58) we have that the accuracy in the world frame will be

$$A_w = \sqrt{\frac{N_G}{2E} \left(\frac{r^2}{\beta_2^2} + \frac{c^2}{\beta_1^2} \right)} . \tag{6.68}$$

□

6.4 Noise Ambiguity

The discussion in Example 6.2 and also Sect. 6.3 showed that the *a posteriori* p.d.f. can be partitioned into two classes, near $\widehat{\boldsymbol{\xi}}$ and not near $\widehat{\boldsymbol{\xi}}$. The p.d.f. near $\widehat{\boldsymbol{\xi}}$ gives an indication of the accuracy of the system, particularly for high signal to noise ratios, the region not near $\widehat{\boldsymbol{\xi}}$ gives an indication of how likely we are to mistake a noise peak for the signal peak i.e. the noise ambiguity.

Woodward [128] defines noise ambiguity (for one-dimensional radar systems) as the fraction of posterior probability which is disconnected from the region near the optimal value.

Denote the Woodward noise ambiguity as P_A. It is difficult to provide a mathematical definition of P_A as the signal and noise are entangled, making it difficult to formally separate the two. However, following Woodward [128] let us proceed as follows. Separate $v(\boldsymbol{w})$ into signal and noise components v_s and v_n and then define the following quantities (based on (6.54))

$$P_s(\xi) = \frac{p(\xi)\exp\left(-\frac{1}{N_g} < |v_s(w) - u(w,\xi)|^2 >\right)}{\int \mathrm{d}\mu\, p(\mu)\exp\left(-\frac{1}{N_g} < (v(w) - u(w,\mu))^2 >\right)} \tag{6.69}$$

and

$$P_n(\xi|v) = \frac{p(\xi)\exp\left(-\frac{1}{N_g} < |v_n(w) - u(w,\xi)|^2 >\right)}{\int \mathrm{d}\mu\, p(\mu)\exp\left(-\frac{1}{N_g} < (v(w) - u(w,\mu))^2 >\right)}. \tag{6.70}$$

P_s is can be thought of loosely as that part of $p_{\xi|v}$ due to the signal while P_n, similarly is that part of $p_{\xi|v}$ due to the noise. Then we will have, for the case of $p(\xi)$ uniform,

$$P_A \approx \frac{\int \mathrm{d}\xi\, \mathcal{E}\{P_n(\xi)\}}{\int \mathrm{d}\xi\, P_s(\xi) + \int \mathrm{d}\xi\, \mathcal{E}\{P_n(\xi)\}} \tag{6.71}$$

(i.e. the area of the *a posteriori* p.d.f. due to the noise alone, divided by the area of the *a posteriori* p.d.f. due to the signal plus noise).

Woodward estimated P_A as a function of R for a pulsed ranging system with *a priori* distribution of ranges which is uniform over a range T_u. The result is

$$P_A = \frac{T_u R\beta}{T_u R\beta + \exp\left(\frac{R}{2}\right)}. \tag{6.72}$$

A plot of this function for the $T_u\beta = 5000$ is reproduced in Fig. 6.4.

As expected, when R is small the ambiguity is very large, whereas for large R the ambiguity is very small. One feature of the graph pointed out by Woodward is the pronounced threshold. For R less than 25, there is little usable information as the ambiguity is very high. For R greater than this value, the ambiguity drops rapidly to zero and it is possible to make meaningful ranging measurements.

The difficulty with Woodward's definition of ambiguity is that the concept of 'disconnected from the region near the optimal value' is hard to define in a precise mathematical fashion. Accordingly the following definition is proposed. Let the generalised bandwidth product for a positioning system be defined as

$$\beta_n = \sqrt{\det(\boldsymbol{C})} \tag{6.73}$$

where $\boldsymbol{C}$ is given by (6.55). Then define M_g as

$$M_g = U_n\beta_n \tag{6.74}$$

where, as usual, U_n is the region of uncertainty in the communications frame. The ambiguity is defined as

$$A_g(R) = P_E \tag{6.75}$$

where P_E is as in (6.31) and

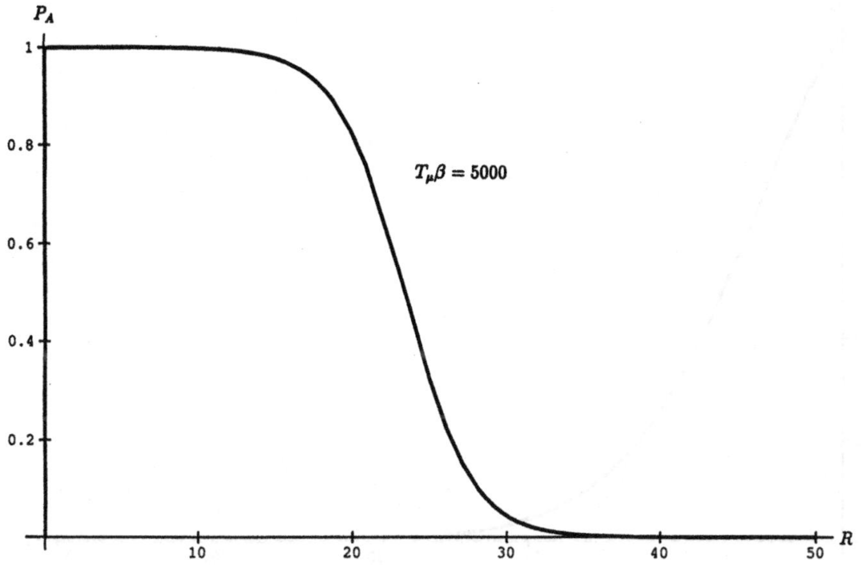

Fig. 6.4. Plot of Noise Ambiguity (Woodward) as a function of R

$$R = \frac{2S_pT_g}{N_g} \tag{6.76}$$

and M is set to M_g when calculating P_E. Note that in many cases (6.31) and Fig. 6.2 can be used to calculate $A_g(R)$.

This definition assumes an n-dimensional channel which is n-ways separable (see Example 6.3). The case of non-separable channels is discussed in Sect. 6.5. As well, the *a priori* object p.d.f. is assumed to be uniformly distributed over a volume U_n in the communications frame. This allows compatibility with the definition of calibration performance (see Sect. 6.6), but if necessary, it should not be too difficult to remove this restriction.

The justification for this definition of ambiguity is as follows. Section 6.1 showed that it is possible to view a positioning system as a communications system which is sending a series of messages about where the object is. These messages are sent by transmitting a waveform, one for each message. The best coding technique will assign an approximately orthogonal waveform for each message. However there will be a limited number of orthogonal messages. The number of allowable orthogonal messages will be the available waveform 'space' divided by the 'width' of the waveforms. Here the 'width' is interpreted to be the reciprocal generalised bandwidth $\frac{1}{\beta_g}$ (see (6.73) and (6.61)) and the available space is U_n. Accordingly, M_g is the estimate of the number of orthogonal messages, and we have a model of the positioning system as a device that sends

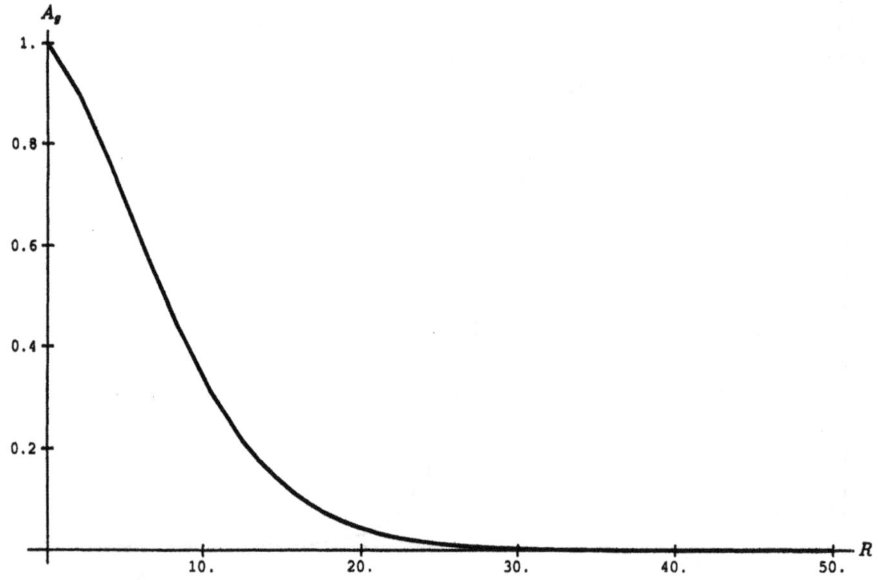

Fig. 6.5. Plot of Noise Ambiguity as a function of noise.

one message out of a possible M_g. Now we can see that ambiguity, as defined by this method, is the probability of receiving the wrong message.

Example 6.4

Consider a ranging system which has a value of β of 1 MHz and a range uncertainty of 5 milliseconds. In this case $M_g = 5000$. The noise ambiguity A_g has been plotted as a function of R in Fig. 6.5. As can be seen the result is qualitatively similar to Fig. 6.4 although the threshold is not as well defined. As will be shown below, the threshold does become well defined when M_g becomes very large. □

Example 6.5 Consider (6.31), for the case of a one-dimensional time-delay positioning system. The value of A_g in the limit as $M_g \to \infty$ can be established using the well known result for the probability of error of a communications channel [122].

$$\lim_{M_g \to \infty} A_g(R) = \begin{cases} 1 & \text{if } R_B < 2\log 2 \\ 0 & \text{if } R_B > 2\log 2. \end{cases} \tag{6.77}$$

where

$$R_B = \frac{2S_pT_B}{N_0}, \tag{6.78}$$

and $T_B \equiv \frac{T}{\log_2 M_g}$ is the observation time per bit. There is obviously a distinct threshold for this case at $R_B = 2\log 2$. □

6.5 Signal Ambiguity

So far this monograph has discussed physical ambiguity and noise ambiguity. Another important form of ambiguity is signal ambiguity. This can occur when a system cannot distinguish between a signal and a translated version of that signal. For example a repetitive waveform will have autocorrelation peaks occurring at the repetition rate of the waveform. A positioning system using such a waveform for ranging measurements will be unable to determine which repetition of the waveform produced a particular echo.

Another form of signal ambiguity can occur if two or more objects are being measured simultaneously. For example, if two objects are close together then it may not be possible to distinguish between them. The ability to distinguish simultaneously between two or more objects will be called the *close resolution*.

Because the important forms of physical coding involve translating the waveform, the optimal receiver (Sect. 6.2) will in most cases perform the correlation operation $< v(w)u(w-\xi) >$.

As we are only concerned with the properties of the signal, we can discuss signal ambiguity in terms of the function

$$R_g(\xi) =< u(w)u(w-\xi) > . \tag{6.79}$$

For most practical waveforms, this autocorrelation function can be divided into two parts: the main peak and the sidelobes. This subdivision is demonstrated for a typical one-dimensional auto-correlation function in Fig. 6.6.

The width of the main peak and the height of the sidelobes are important for three reasons, the first is the accuracy, the second is resolution and the third is performance. This is discussed below.

Accuracy Equation (6.68) shows that, for high signal-to-noise ratios the system accuracy depends inversely on the r.m.s. widths of the power spectrum. But the Fourier transform of the power spectrum is the auto-correlation function, so that the accuracy will depend directly on the r.m.s. widths of auto-correlation function. The narrower the main peak, the greater the system accuracy. However if the sidelobes are high, the noise ambiguity will be large and the overall system accuracy cannot be determined from an examination of the width of the peak alone.

The effect of sidelobes on noise ambiguity can be seen from the following argument. The noise ambiguity is calculated by considering M_g different signals. In this case each signal will be a simple translation so that the elements of $\boldsymbol{R}$ will be proportional to samples of the auto-correlation function, for example

$$R(\boldsymbol{\xi}_i - \boldsymbol{\xi}_j) = E\rho_{ij} \tag{6.80}$$

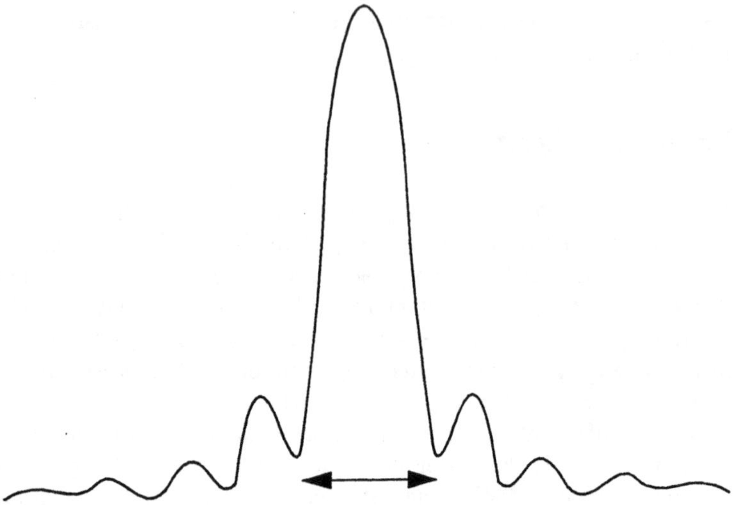

Fig. 6.6. Typical Auto-correlation Function

where E is the total energy in the waveform and $R(\cdot)$ is the correlation function. Now the derivative probability of error, P_E, with respect to any of the $\rho_{ij}, i \neq j$, is positive [122]. Accordingly, decreasing the height of the sidelobes outside the main peak will decrease P_E and so the noise ambiguity.

Resolution The size of the main peak will be an important factor in determining the close resolution performance of a system. Close resolution is a measure of the ability to differentiate closely spaced objects. The most common definition of resolution, the two-point resolution [40], is based on the ability of a system to resolve two closely spaced points. For many years the limit of this form of resolution was thought to be the Rayleigh criterion which states that the two points can be separated 'if the centre of the diffraction pattern corresponding to one point coincides with the first diffraction minimum for the other point' [131]. Accordingly, the width of the main peak can be used as a guide to close resolution performance.

However it should not be used as an absolute indicator. It has been found that the Rayleigh criterion does not provide a limit to the resolution of systems. This phenomenon is referred to variously as super directivity [19], super gain [27], super resolution, ultra resolution [18] and high resolution [86]. They can all be thought of as performance enhancement schemes. Performance enhancement schemes will be discussed in more detail at the end of this chapter.

The size of the sidelobes will also have an effect on resolution, but in this case the resolution of targets that are far apart. If one target has a much higher signal to noise ratio than another, then the sidelobe of the stronger target can obscure the main peak of the weaker target. This is often called the near far problem because the differences in signal strength result from the stronger target being much closer. As for the case of noise ambiguity, the optimal auto-correlation function will be the one that has the smallest sidelobes.

Performance The third factor that the width of the peak affects is the overall system performance. This is because the number of independent messages that can be sent depends on the width of the peak (see Sect. 6.4). Accordingly, to a first order, the smaller the size of the main peak, the greater the system performance.

In a finite bandwidth system, it is not possible to alter the size of the main peak without affecting the height of the sidelobes which in turn will increase the noise ambiguity and so reduce the performance. This tradeoff is the key to waveform coding.

From the above discussion we can now visualise the autocorrelation function of the optimum waveform for a positioning system. It should have zero sidelobes and an infinitely narrow main peak. Such a waveform would require infinite bandwidth. It is shown in Sect. 6.6 that such a waveform can approach the Shannon limit on performance. The method of determining the optimum waveform when there are constraints will be discussed in Sect. 6.8.2.

The preceding discussion also shows that by examining the auto-correlation function, an experienced designer can formulate a qualitative feel for the system accuracy, resolution, and performance. One sort of system for which this analysis can be done is a positioning system which simultaneously measures range and range rate, such as Moving Target Indication radar [107]. In this case the range is coded as time delay and the range rate as frequency shift. Here the normal auto-correlation function cannot be used because in this case w is one-dimensional (time) but ξ is two-dimensional (time and frequency). Accordingly the waveform cannot be written as $u(w - \xi)$. In this case, we know from (6.54) that the optimal receiver will perform the operation $< |v(w) - u(w, \xi)|^2 >$. If the *a priori* distribution is uniform, the optimum receiver will seek to minimise this function.

Suppose that our waveform is $u(t)$. Let us write this in complex notation [33] as $\gamma(t)$, where

$$\gamma(t) = (u(t) + \jmath\mathcal{H}[u(t)])e^{-\jmath 2\pi f_0} \tag{6.81}$$

where f_0 is the centre frequency of the signal and $\mathcal{H}$ denotes the Hilbert Transform. The function $\gamma(t)$ is the complex envelope of $u(t)$.

Now a shift by τ in time and a shift by ν in frequency will transform $\gamma(t)$ to $\gamma(t+\tau)\exp[\jmath 2\pi(f_0\tau + \nu t)]$. Consider $< |v(\boldsymbol{w}) - u(\boldsymbol{w}, \boldsymbol{\xi})|^2 >$, in the case of zero noise the receiver should seek to minimise

$$< |u(t) - u_{\tau,\nu}|^2 >= 2 < u(t)^2 > -2 < u(t)u^*_{\tau,\nu} > . \tag{6.82}$$

Then [33]

$$< |u(t) - u_{\tau,\nu}|^2 >=< |\gamma(t)| >^2 - \mathcal{R}(\chi(\tau,\nu)) \tag{6.83}$$

where χ is defined as

$$\chi(\tau,\nu) = e^{-2\pi f_0 \tau} \int_{-\infty}^{\infty} \gamma(t)\gamma^*(t+\tau)e^{-2\pi\nu t}\,dt \tag{6.84}$$

and $\mathcal{R}$ denotes the real part of the expression.

The squared modulus of χ is called the radar ambiguity function [97, 108]. It can also be thought of as a time-frequency distribution [16]. The above derivation shows that the function plays the same role as the auto-correlation function discussed earlier. It can be used in a similar fashion to the auto-correlation to judge the accuracy, performance and resolution of a radar [97]. In addition, it can be used to determine the relative accuracy in the time and frequency domain. In radar and other positioning systems there is often a trade-off between doppler and ranging performance. The radar ambiguity function can be used to gain an understanding of this trade-off. But there is another way to look at this issue.

For an n-dimensional positioning system with capacity C, the sum of the individual information transfer between each of the co-ordinates must be less than the total information transfer which in turn must be less than the capacity times the observation time:

$$\sum_{i=1}^{n} I(x_i; \hat{x}_i) \le I(\boldsymbol{x}; \hat{\boldsymbol{x}}) \le CT. \tag{6.85}$$

More specifically, for a system that measures the range r and the range rate r', we have

$$I(r;\hat{r}) + I(r';\hat{r}') \le I(r,r';\hat{r},\hat{r}') \le CT. \tag{6.86}$$

This shows unequivocally the trade-off between the range performance and the range-rate performance of the system. It is a true uncertainty relationship in the sense that it establishes a limit on the simultaneous measurement of range and range rate.

6.6 Performance of Coding Schemes

Consider a system which measures the position of a series of objects. The question is, given a particular coding scheme, what is the performance of the system?

If the source statistics and measurement constraint are known, then it is possible to calculate the performance of the system for a particular coding scheme, but such an answer would be specific to the source statistics and measurement constraint. Information on these quantities may not be completely available or a licensing authority that is making a comparison of systems may be interested in these matters, as they relate more to system usage than capability.

In a typical communications system this problem is handled by making reasonable assumptions about the source statistics in order to calculate a bit rate and an associated error probability (see Sect. 6.1). The normal assumptions for communications systems are that each source symbol is independent, each letter of the source alphabet is independent and only one symbol is sent at a time. This allows the publication of a system bit rate which can be readily used to compare different systems and coding schemes. If a user chooses to utilise a communication system by sending data at a much lower rate than the data rate, then that is of no concern to the licensing authorities who allocate on the basis of system capability rather than system usage.

The aim here is to provide reasonable assumptions concerning the measurement strategy and source statistics that will allow a meaningful comparative performance figure for different coding schemes. This comparative figure will be known as the *calibration performance.*

In order to make our definition we have to consider the assumptions we need to make about source statistics and measurement strategy.

Source Statistics There are two assumptions necessary concerning the source statistics. The first is the correlation between individual measurements. The performance will be best when each measurement is independent, so it will be assumed that each measurement is independent. The second aspect of the source statistics concerns the nature of the *a priori* distribution of the objects. This constraint could be in terms of specifying a known area that the objects must lie within, or specifying a known positional p.d.f. In the definition of calibration performance developed here, the objects are constrained to be uniformly distributed over a particular domain. This requires less knowledge on the part of systems engineer than the assumption of a particular p.d.f. It is easy to adapt the definition presented below to the case when the p.d.f. is known. Note that the p.d.f. is assumed uniform in the communications frame, not the world frame. The reason for this is that all the calculations concerning coding will be carried out in the communications frame.

Measurement Strategy There are two features of the measurement strategy that need to be considered; measurement order and measurement accuracy. Just as in calculating the bit rate for a communications system, where there is only one transmitter, for a positioning system it will be assumed that only one measurement will be taken at a time. Our assumption concerning independence of the measurements makes the selection of measurement order simple: the order of measurement makes no difference. The second aspect to be considered is measurement accuracy. Sometimes, there will be a requirement that each measurement must be made to a specific accuracy which means that each measurement must allow sufficient integration time to achieve this accuracy. This corresponds to a requirement in a communications system to achieve a particular error rate. Hence it is necessary to include an accuracy in our definition of calibration performance.

Given these assumptions we are now able to make a definition of calibration performance:

The *calibration performance* for a positioning system operating on a particular domain of interest with a specified r.m.s. accuracy will be arrived at by calculating the information performance (see Sect. 2) assuming that each measurement is independent, only one measurement is made at a time, the positional p.d.f. for each object is uniformly distributed (in the communications frame) over the volume of interest, and a specified accuracy is achieved. The specified accuracy may vary over the domain of interest in which case it may be simpler to specify the signal power, bandwidths and integration time and use these parameters to calculate A_w (see (6.58)).

For most systems the calibration performance will vary depending on the specified value of accuracy. If there is a value of accuracy which gives the largest value of calibration performance, then this value of performance will be known as the *maximum calibration performance.*

In order to calculate the calibration performance it is necessary to evaluate the average mutual information between the parameter being sent, $\boldsymbol{\xi}$ and the received signal waveform $\boldsymbol{v}$. We have that

$$I(\boldsymbol{\Xi}; \boldsymbol{v}) = H(\boldsymbol{\Xi}) - H(\boldsymbol{\Xi}|\boldsymbol{v}). \tag{6.87}$$

In the case of calibration performance $H(\boldsymbol{\Xi})$ is easily calculated, as it will be the entropy of a uniform distribution. In general, $H(\boldsymbol{\Xi}|\boldsymbol{v})$ will be very difficult to calculate, though in some simple cases it is possible to make progress.

Example 6.6

Consider a pulsed ranging system, with *a priori* range distribution of T_u, and a centre frequency of f_0. Denote the time delay as τ, the signal as u, the noise as n and the received waveform as $v = u+n$. Suppose the actual transmitted value of time delay is τ_0. The noise is assumed to be white and gaussian. In calculating the calibration performance we assume that the *a priori* distribution is uniform, so

$$H(\tau) = \log(T_u). \tag{6.88}$$

It is not possible to analytically determine $H(\tau|v)$ over the entire range of R, but it is possible to make a calculation when R is very small or R is very large. Woodward [129] shows for the case of R small (i.e. the noise ambiguity is close to 1) and ignoring the information loss due to smoothing out of the fine structure of the signal i.e. envelope detection, the information gain is given by:

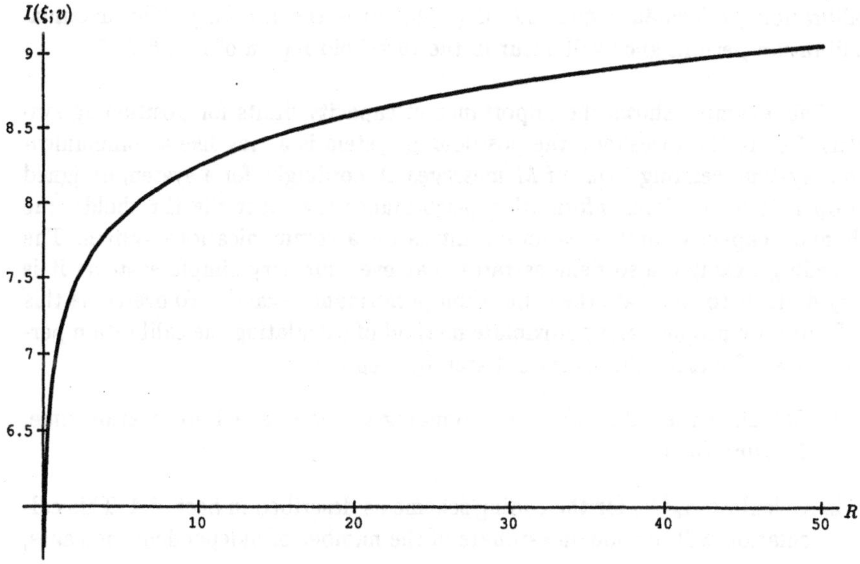

Fig. 6.7. Information Gain as a function of R.

$$I(\xi; v) \simeq \frac{1}{2}R. \tag{6.89}$$

When the value of R is large so that the noise ambiguity is small then $\xi|v$ will be gaussian distributed with zero mean and variance given by (6.57). Hence in this case

$$I(\xi; v) = \log(T_u) - \frac{1}{2}\log\left(\frac{2\pi e}{R\beta^2}\right). \tag{6.90}$$

Figure 6.7 shows a plot of $I(\xi; v)$ as a function of R, for the second case. The value of $T_u\beta$ is 5000.

These two equations show that initially the information gain rises in a linear fashion, until the threshold region (noise ambiguity is close to zero), when the gain rises as a logarithmic function of T. This also gives an indication of how best to achieve optimal performance from a positioning system (i.e. to avoid integrating after the noise ambiguity is close to zero). If this is unavoidable, then a frequency multiplexing scheme such as described by Hurst [48] can be employed. Of course, having multiple targets also increases the performance of a positioning system, but more work needs to be done to analyse this case.

Equation (6.90) can be used to calculate the calibration performance of the system. If R is large enough for this equation to be valid, (6.58) will provide a useful approximation to the system accuracy (see (6.68) to see how A_w is a function of $R = \frac{2E}{N_0}$). Accordingly, by choosing a particular integration time we effectively fix R. Then dividing (6.90) by the integration time gives the calibration performance and use of (6.58) gives the accuracy. The maximal calibration performance will occur in the threshold region of Fig. 6.7. □

This example shows the importance of capacity limits for positioning systems. Before the threshold, the positioning system is acting like a communications system, sending 1 out of M messages. Accordingly, for a system designed to operate for optimal information performance (i.e. near the threshold), the Shannon capacity limit is as important as for a communications system. The preceding example also demonstrates that even for very simple systems it is very difficult to calculate the calibration performance exactly. To overcome this difficulty, we propose an approximate method of calculating the calibration performance. The algorithm is set out step by step below:

1. Calculate the value of R corresponding to the selected observation time, T, using (6.76).

2. Calculate $A_g(R)$ for the coding scheme as described in Sect. 6.4. This calculation will include an estimate of the number of independent messages, M_g.

3. If there were no error, the total number of bits of information transferred would be $\log_2 M_g$. However the noise ambiguity error will cause a reduction in this amount of information. Assuming that an error is equally likely to be associated with any of the messages, it can be shown that

$$I_g(R) = \log_2 M_g + (1 - A_g(R)) \log_2 (1 - A_g(R)) + A_g(R) \log_2 \left(\frac{A_g(R)}{M - 1} \right) \tag{6.91}$$

   where I_g is the average mutual information estimated using the noise ambiguity approach [5, p113]. The first term on the right hand side is the message entropy and the next two terms represent the loss due to errors.

4. When the noise ambiguity is low, the *a posteriori* p.d.f. $p(\xi|v)$ will be approximately gaussian, with the covariance matrix determined by (6.61). The average mutual information in this case will be $H(\xi) - H(\xi|v)$ or

$$I_a(R) = \log U_n - \frac{1}{2} \log \left(\left(\frac{2\pi e}{R} \right)^n \det(C^{-1}) \right) \tag{6.92}$$

   where I_a is the average mutual information.

5. When the noise ambiguity is high, then I_g should be an accurate estimate, when it is low then I_a should be accurate. Accordingly a reasonable weighting between the two should be $A_g(R)I_g(R) + (1 - A_g(R))I_a(R)$ and our estimate of the calibration performance will be

$$P \simeq \frac{A_g(R)I_g(R) + (1 - A_g(R))I_a(R)}{T}. \tag{6.93}$$

6. Calculate A_p, the calibration accuracy, using (6.58) and (6.61), noting that $R = \frac{2E}{N_g}$:
$$A_p \simeq A_g(R)A_u + (1 - A_g(R))A_w \tag{6.94}$$
where A_u is the r.m.s. error assuming no knowledge of the object's position other than the positional p.d.f.

7. If needed, the calibration performance can be calculated as a function of A_p, because both P and A_p are functions of R. The maximum value of this function will give the maximum calibration performance.

It is worth noting that many systems use a dual approach to the signal processing involved in waveform decoding. Initially, when little is known of the object's position, a bank of correlators or a sequential acquisition approach is used to work out the approximate object location. Once this is known, the system will switch to a second mode, such as 'tau-dither' tracking of the correlation peak. During the acquisition phase, the noise ambiguity approach seems to be a good model of system performance (with M_g equal to the number of correlators), while during the tracking mode the accuracy method seems more appropriate. Accordingly, for many systems, it is easy to work out when to use I_g and when to use I_a.

Some caution should be used when applying these formulae. In particular the noise ambiguity method of calculating the average mutual information transfer will tend to give an overestimate for the reasons discussed in Sect. 6.1. If the algorithm is to be used for waveform selection then the method is probably adequate. If it is to be used for estimating acquisition time [26] for actual system operation then appropriate corrections should be made to account for incoherent detection [5, p154], quantisation errors, and other factors.

Example 6.7

Consider a one-dimensional ranging system, with $U_n = 511\ \mu s$, $\beta = 1\ \mu s$, $\frac{2S_p}{N_0} = 0.005$, and $M = 511$. A plot of the calibration performance is shown in Fig. 6.8 and the accuracy in Fig. 6.9. The maximum value of the calibration performance occurs at about 1800 milliseconds. The accuracy at

this point is about 70 metres. Normally, there would be further integration of the signal to obtain a superior accuracy (i.e. the maximum calibration performance is not a measure of the optimal system accuracy).

□

It is interesting to use this method to consider how well a positioning system performs when the bandwidth becomes infinite. If the bandwidth of the channel is allowed to become infinite then the performance of a positioning system can approach the Shannon Limit. This result can be established as follows: in Sect. 6.4 the noise ambiguity for a channel with an infinite number of messages was calculated. The result was
$$\lim_{M_g \to \infty} A_g(R) = \begin{cases} 1 & \text{if } R_B < 2\log 2 \\ 0 & \text{if } R_B > 2\log 2. \end{cases} \tag{6.95}$$

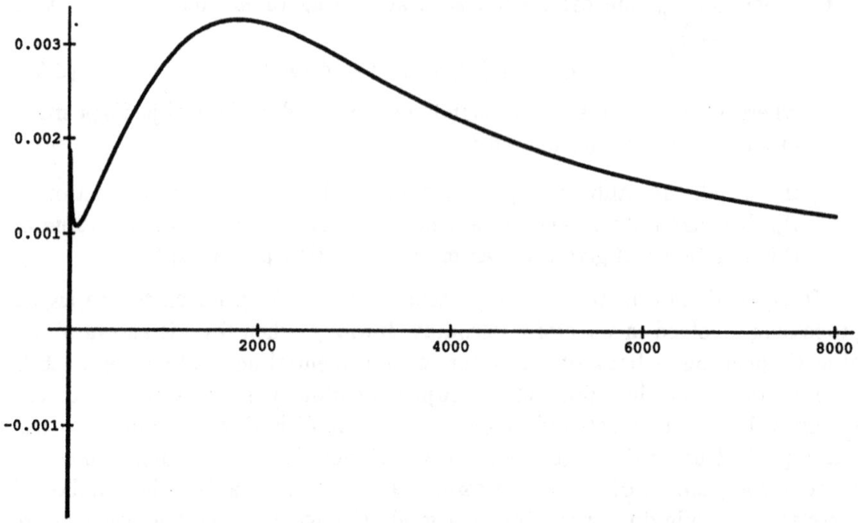

Fig. 6.8. Plot of Calibration Performance vs time (in milliseconds)

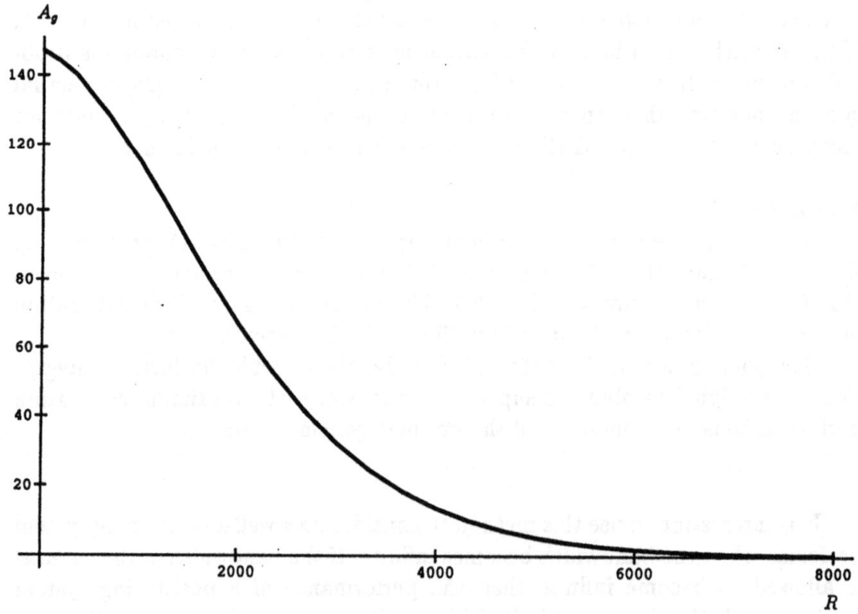

Fig. 6.9. Plot of accuracy vs time (in milliseconds)

Remembering that $R_B = \frac{2S_pT_B}{N_g}$, we have that for error-free transmission

$$\frac{1}{T_B} < \frac{S_p}{N_0 \log_2 2}. \tag{6.96}$$

T_B is the time to measure one bit, so that the data rate can approach $\frac{S_p}{N_0 \log_2}$ bits per second. As the number of messages is infinite, there will be no quantisation error, so this data rate will equal to the calibration performance of the positioning system.

Now let us compare this calibration performance with the theoretical capacity of a one-dimensional channel when the bandwidth approaches infinity. The capacity of the channel will be given by [6]

$$C = B \log_2 \left[1 + \frac{S_p}{N_0 B}\right]. \tag{6.97}$$

Now consider the limit as B goes to infinity.

$$\lim_{B \to \infty} C = \left(\frac{S_p}{N_0}\right)\left(\frac{N_0 B}{S_p}\right) \log_2 \left[1 + \frac{S_p}{N_0 B}\right], \tag{6.98}$$

so

$$\lim_{B \to \infty} C = \frac{S_p}{N_g} \log_2 \left[1 + \frac{S_p}{N_0 B}\right]^{\frac{N_0 B}{S_p}}. \tag{6.99}$$

Given that [1] $\lim_{a \to \infty}(1 + \frac{1}{a})^a = e$ we have

$$\lim_{B \to \infty} C = \frac{S_p}{N_0} \log_2 e \tag{6.100}$$

or

$$\lim_{B \to \infty} C = \frac{S_p}{N_0} \log 2. \tag{6.101}$$

Accordingly, by considering (6.96) we can see that in the limit, the performance of a positioning system can approach the ideal. This was first pointed out by Woodward [128]

It is worth noting that although the bandwidth becomes infinite and the performance of the system approaches the Shannon limit the accuracy of the system does not become infinite. This is despite the fact that the accuracy calculated using (6.57) and (6.61) does become infinite. The reason for this is that as the bandwith grows, so does the number of possible messages. Although the correct message will have very high accuracy, the chances are high, given the large number of messages, that one of the other messages will be mistaken for the correct one. Accordingly the noise ambiguity does not drop to a sufficiently low level to allow proper application of (6.57). Another way to come to this conclusion would be to observe that if the accuracy were infinite the information transfer would be infinite, which would be greater than the Shannon limit for a channel with infinite bandwidth but finite signal power.

[6]The following development follows Viterbi [122].

6.7 Performance Enhancement Schemes

In Chap. 5, it was pointed out that if the source information rate is less than the system capacity then the performance will be bounded by the source information rate. Indeed, in such cases, the performance can be exactly equal to the source information rate. This can be seen by the following simple example.

Example 6.8

Consider a one-dimensional positioning system with a finite bandwidth which is trying to determine the location and velocity of an object by measuring the time delay and doppler shift of a series of pulses. A transmitter emits a pulse, the pulse echoes from the object and a receiver that is collocated with the transmitter measures the round trip time.

From (6.92) we know that the long-term information gain for such a system will be

$$I_a(R) = \log U_n - \frac{1}{2}\log((2\pi e)^n \det(C)). \tag{6.102}$$

Given that R is a linear function of time and that C and U_n are independent of time, the rate of change of information gain will be

$$\frac{\mathrm{d}I_a(R)}{\mathrm{d}t} = \frac{n}{2t}. \tag{6.103}$$

In this case, $n = 2$, as we are estimating position and velocity, so that

$$\frac{\mathrm{d}I_a(R)}{\mathrm{d}t} = \frac{1}{t}, \tag{6.104}$$

which agrees exactly with the absolute source information rate of a constant velocity object (see (5.52)).

□

Accordingly, after the initial acquisition[7], the information performance of the system is independent of the classical resolution limit and is equal to the long term source information rate. Furthermore, this long term information rate approaches zero (i.e. is small). Given that many positioning systems measure constant velocity objects, this example shows that for such systems the performance will be bounded by the very low source information rate of such objects, unless an innovative performance enhancement scheme is devised. And indeed, many such schemes have been proposed.

Using the characterisation of positioning systems presented in this monograph, it is possible to clarify the mechanisms involved in performance enhancement schemes. This will be demonstrated by considering four different performance enhancement schemes, namely resolution enhancement, directional trade-off, independent source coding and Synthetic Aperture Radar (SAR).

---

[7]During acquisition the information rate can be much higher than the long term source information rate but will still be bounded by the channel capacity.

Resolution Enhancement Resolution enhancement includes analytic continuation and super gain techniques. Both will be discussed here.

Analytic continuation is a technique of use in imaging systems [40, 18]. In essence it attempts to improve the resolution of an image by sampling the image in the pass band and extrapolating to outside the pass band. It is based on the fact that a Fourier transform of a spatially bounded object is analytic and so can be determined everywhere by *exactly* determining the transform in a finite region.

In essence this is a post-processing technique. In terms of our model, this technique does not perform any extra coding, apart from the original physical coding. Cox and Shepherd [18] suggested that the performance of the analytic continuation can be judged by a trade-off between the number of degrees of freedom arising from the system capacity. This may be correct, but in many cases the correct number of degrees of freedom should be derived from the source information rate which will often be much lower than the system capacity, particularly if there is long term integration of the image to improve the signal to noise ratio.

There are a variety of other techniques which use *a priori* information about the objects (e.g the object is of finite extent, it is stationary, there are only two objects, etc.) in order to make maximum use of the incoming information. Examples include super gain or super directivity [19]. Such estimation techniques do not increase the source information rate because a proper evaluation of the source information rate will take into account the *a priori* information. Accordingly these techniques are only providing a different view of the incoming data. Of course this different view may be useful, indeed it can increase the overall information performance. Accordingly it is proposed that such techniques be called resolution enhancement.

Directional Trade-off Another technique which is classified as a super-resolution technique increases resolution in one direction by using spare bandwith in an orthogonal direction [76, 77]. In essence this technique works by taking information that lies outside the multi-dimensional pass band of the system and recoding it so that it lies within the pass band. At the other end, the information is decoded and reconstituted to yield a higher resolution version of the original image. Accordingly this method achieves a higher rate of information gain by increasing the source information rate in the pass band. As this increase is achieved by changing the physical coding (e.g. interposing a special lens in the transmission path) it is suggested that techniques of this kind be called *augmented physical coding*.

Independent Source Coding Suppose the one-dimensional positioning system described above is applied to the problem of tracking vehicles in urban areas. Such vehicles will have a relatively low source information rate. This suggests the possibility of tracking more than one vehicle at a time using the same bandwidth. In fact there is a scheme that takes advantage of the low source information rate of vehicles [132]. If the system transmits a train of pulses then because each successive pulse will be delayed by a similar amount, the spectrum becomes combed. By slightly offsetting the centre frequency of each vehicle it is possible to interleave the combs without interference. Although it does not improve the resolution of each vehicle measurement, it does significantly increase the system information rate by increasing the number of independent sources. It is an example of augmented source coding which takes advantage of the low source information rate of the objects.

Synthetic Aperture Radar (SAR) In this case, a moving antenna is used to synthesize an aperture, so giving a much higher resolution than is possible from the aperture of the original antenna [65]. The effect of moving the antenna is equivalent to producing a synthesized antenna with a much larger aperture. This increases the spatial bandwidth of the system. A SAR system achieves increased resolution by trading off the temporal degrees of freedom in favour of the spatial degrees of freedom. Accordingly the system capacity is not altered.

However the information performance is radically improved. In communications systems terminology, this increase is achieved because the number of near orthogonal messages is greatly increased (due to the increased spatial bandwidth) without affecting the received signal energy. Thus instead of sending a few signals with a long integration time (and so being bounded by the long term low absolute source information rate), the SAR will send many signals with a much shorter integration time. The shorter integration time means more chance of making an error as to which signal has been sent, but the larger number of signals means a much greater nett bit rate. This technique achieves the increase information performance by changing the physical configuration of transmitter and receiver, so is another example of augmented physical coding.

The above shows that the described performance enhancement schemes can be divided into three categories: resolution enhancement, augmented source coding and augmented physical coding. Analytic continuation and super-directivity are examples of resolution enhancement. Directional trade-off and SAR are examples of augmented physical coding and independent source coding is an example of augmented source coding.

The augmented physical coding and source coding examples demonstrate that if the source information rate is considerably less than the channel capacity then there could be scope for performance enhancement. The extent of the spare capacity can be quantified in the following manner. If the channel has a capacity C, and the measurements start at time $t = 0$, then after a period T, the unused capacity available for additional encoding is

$$C_e = CT - \int_0^T s(t)\,\mathrm{d}t\,. \tag{6.105}$$

This is particularly important for systems measuring objects which can have low long term source information rates where there should be great scope for augmented coding schemes. Of course it will be up to the ingenuity of the system designer to achieve this benefit. However the examples of SAR, independent source coding, directional trade-off and other techniques indicate that it is possible in a wide variety of systems. The recent discoveries in Chaos theory should make it easier for systems engineers to estimate the source information rate. This innovation, together with the method of characterising positioning systems presented in this monograph, should increase our understanding of the complex mechanisms involved in the performance of positioning/imaging systems and so assist in the invention of new performance enhancement schemes.

6.8 Other Areas of Investigation

The topics discussed in this chapter invite further research in a number of areas. Two possible extensions of this work will be discussed here: realistic channels and waveform selection.

6.8.1 Realistic Channels

Most of the work in this book is based on the assumption of additive, white, gaussian noise (AWGN) channels. Often real world systems do not neatly fit this assumption (either in the temporal or spatial dimensions of the channel). However there are strong reasons for first analysing channels as if they were AWGN [5]:

- often, an AWGN channel is a reasonable fit to reality,
- AWGN channels are mathematically tractable so it is possible to gain a deeper insight into the underlying processes, and
- often rankings of systems/techniques done under the AWGN assumption will hold under more realistic situations.

Given all this, it seems reasonable that AWGN channels are a good initial approach for the method of characterizing positioning systems described here. However, to make the work applicable to a wider range of problems, it will be necessary to extend the analysis to cover some of the system degradations which are due to other than AWGN. These include non-linearities, phase/timing errors, spatially correlated noise, inter-channel interference and multi-path fading.

This list (and it can be much longer) appears daunting. However one of the features of this work is that it allows the characterisation of positioning systems in similar terms to a conventional communications system, for example by defining bit rate, probability of error, information performance, capacity etc. This means that the enormous body of work done in this area (see Benedetto [5] for an overview) can be applied, with suitable modification, to the problems of real world positioning systems. Of the areas of attack, probably the most important area is multi-path fading, where the techniques and knowledge gathered by the mobile communications fraternity should be particularly applicable [73].

6.8.2 Wave-Form Selection

The problem of waveform selection has a long history. One mathematical basis was established by Woodward [128]. Much of the work in this area is described by Rihaczek [97]. More recent work has looked innovatively at the performance of specific coding methods (e.g. [17, 39]).

The previous section showed that it is possible to estimate the calibration performance of a particular coding method. Accordingly we propose a method for evaluating different waveforms by comparing the calibration performance of different waveforms. This could be applied to the important problem of finding the optimal waveform subject to a set of constraints. The most common constraints will be on signal energy, observation time, peak signal power and bandwidth. As well some of the constraints could be based on the partitioning of the average mutual information between different variables (e.g. time delay and Doppler shift).. Once the constraints have been set up, it should be possible to use standard numerical optimisation techniques [94] to obtain a sampled version of the optimal auto-correlation function. Then a realisable waveform could be calculated which satisfies the constraints and has the maximum calibration performance.

It is interesting to compare this technique with the radar ambiguity function of waveform selection. The advantage of the method proposed here is that it may provide a quantitative answer to the question of what is the best waveform. As well, it offers the hope of an automatic method for the design of the optimal auto-correlation function that satisfies the constraints. There are two disadvantages with the method proposed here. The first is that only one object is being considered at a time, whereas the radar ambiguity function allows the consideration of a number of targets [97]. This may prove not to be too serious as the proposed method does account for noise ambiguity and accuracy, which are related to far-target resolution and near-target resolution respectively. The second disadvantage of the proposed method is the very large numerical load for practical waveforms. However this problem should be resolvable by applying more sophisticated numerical techniques. Much more work needs to be done to develop this method.

7. Estimation

The previous chapters discussed waveform decoding and physical decoding. The result of these processes is a calculated position for a remote object. If this process is repeated over a period of time then there will be a sequence of such measurements. As usual the vector of measured positions will be called $\boldsymbol{Y}$ and the actual positions will be $\boldsymbol{X}$.

If the user of the positioning system has *a priori* knowledge concerning the kinematics of the object, it is possible to filter the measurements $\boldsymbol{Y}$ to produce a more accurate estimate of $\boldsymbol{X}$. The basic idea is to produce an accurate as possible estimate, given that the user has some knowledge on how the object should behave.

For stationary systems, the optimal (in a mean-square sense) linear filter is either a Wiener filter [100] or a Kalman filter [50]. There is a fundamental equivalence between the two approaches [100] with the Wiener filter operating in the frequency domain and the Kalman filter operating in the time domain. However, unlike the Wiener filter the Kalman filter can be used for non-stationary systems. Accordingly the Kalman filter (or augmented versions) is the most popular approach for filtering data from positioning systems.

Because of the importance of Kalman filters, and to provide a basis for Sect. 7.3, Sect. 7.1 will provide a brief derivation of the Kalman filter equations. Sect. 7.2 describes the Kalman filter algorithm. Sect. 7.3 shows how, under certain circumstances, it is possible to use the output of the estimation process to calculate the system performance. Sect. 7.1 and Sect 7.2 may be omitted by readers familiar with derivations of the Kalman filter.

7.1 Derivation of Kalman Filter

The derivation of the Kalman filter described here is based on treatments by Sage and Melsa [100].

Suppose the object's state equation satisfies a first order difference equation

$$\boldsymbol{x}_{k+1} = \boldsymbol{A}(k+1,k)\boldsymbol{x}_k + \boldsymbol{\Gamma}(k)\boldsymbol{w}_k \tag{7.1}$$

where the system noise $\boldsymbol{w}$ is a zero-mean white noise process with covariance

$$\mathcal{E}\{\boldsymbol{w}_k\boldsymbol{w}_j^{\mathrm{T}}\} = \boldsymbol{V}_w(k)\delta(k-j). \tag{7.2}$$

Here $\boldsymbol{A}(k+1,k)$ denotes a matrix depending on the $(k+1)$th and kth positions. Similarly $\boldsymbol{\Gamma}(k)$ is a matrix dependent on the kth position.

The measurement model is assumed to be given by

$$\boldsymbol{y}_k = \boldsymbol{H}_o(k)\boldsymbol{x}_k + \boldsymbol{v}_k \tag{7.3}$$

where the measurement noise is assumed to be a zero-mean white-noise process with covariance

$$\mathcal{E}\{\boldsymbol{v}_k\boldsymbol{v}_j^{\mathrm{T}}\} = \boldsymbol{V}_v(k)\delta(k-j). \tag{7.4}$$

It is assumed that $\boldsymbol{w}$ and $\boldsymbol{v}$ are uncorrelated.

Based on the series of measurements $\boldsymbol{Y} = (\boldsymbol{y}_1, \ldots, \boldsymbol{y}_k)$,we desire to produce an improved estimate of some x_j. If j is less than k the estimation process is referred to as *smoothing*, if j is equal to k it is *filtering* and if j is greater than k it is called *prediction*. In this treatment we will be concerned only with filtering, but it is easy to extend the results to the other two cases [100].

It should be noted that the restriction of (7.1) to a first order difference equation does not restrict the system under investigation to first order systems. The restriction can be surmounted by including higher order terms in the state variables. This approach is demonstrated by the example set out below, which is adapted from Bozic [9].

Example 7.1

Suppose a one-dimensional system satisfies the differential equation

$$\frac{\mathrm{d}^2x}{\mathrm{d}t^2} = w(t) \tag{7.5}$$

where w is a time varying acceleration. The appropriate difference equation would be of the form

$$x_k = ax_{k-1} + bx_{k-2} + cw_{k-1}. \tag{7.6}$$

In order to change this equation into a first order equation, use the following transformations

$$x_{1,k} = x_k \tag{7.7}$$

and

$$x_{2,k} = x_{k-1}, \tag{7.8}$$

so that

$$x_{1,k} = ax_{1,k-1} + bx_{2,k-1} + cw_{k-1} \tag{7.9}$$

and

$$x_{2,k} = x_{1,k-1}. \tag{7.10}$$

In vector form, this becomes

$$\boldsymbol{x}_k = \boldsymbol{A}\boldsymbol{x}_{k-1} + \boldsymbol{w} \tag{7.11}$$

where

$$\mathbf{A} = \begin{pmatrix} a & b \\ 1 & 0 \end{pmatrix} \tag{7.12}$$

and

$$\mathbf{w}_k = \begin{pmatrix} w_k \\ 0 \end{pmatrix}. \tag{7.13}$$

□

There are a number of ways that the Kalman filter equations can be derived. In line with our previous derivations of the optimal receiver, the approach taken here is to maximise the *a posteriori* probability. In the treatment presented here, it will be assumed that the $\boldsymbol{w}$, $\boldsymbol{v}$ and the initial condition $\boldsymbol{x}_0$ have gaussian distributions. Although this is a restrictive assumption, it allows a simpler derivation and gives the same result as more general methods.

The maximum *a posteriori* estimate seeks to find the maximum of $p(\boldsymbol{x}_k|\boldsymbol{Y}_k)$ where $\boldsymbol{Y}_k = (\boldsymbol{y}_1, \ldots, \boldsymbol{y}_k)$. The theorem of joint probability can be used to write

$$p(\boldsymbol{x}_k|\boldsymbol{Y}_k) = \frac{p(\boldsymbol{y}_k, \boldsymbol{Y}_k)}{p(\boldsymbol{Y}_k)} \tag{7.14}$$

or, by partially unwrapping $\boldsymbol{Y}_k$,

$$p(\boldsymbol{x}_k|\boldsymbol{Y}_k) = \frac{p(\boldsymbol{x}_k, \boldsymbol{y}_k, \boldsymbol{Y}_{k-1})}{p(\boldsymbol{y}_k, \boldsymbol{Y}_{k-1})}. \tag{7.15}$$

Further application of the joint probability theorem gives

$$p(\boldsymbol{x}_k|\boldsymbol{Y}_k) = \frac{p(\boldsymbol{y}_k|\boldsymbol{x}_k, \boldsymbol{Y}_{k-1})p(\boldsymbol{x}_k|\boldsymbol{Y}_{k-1})p(\boldsymbol{Y}_{k-1})}{p(\boldsymbol{y}_k|\boldsymbol{Y}_{k-1})p(\boldsymbol{Y}_{k-1})} \tag{7.16}$$

or

$$p(\boldsymbol{x}_k|\boldsymbol{Y}_k) = \frac{p(\boldsymbol{y}_k|\boldsymbol{x}_k)p(\boldsymbol{x}_k|\boldsymbol{Y}_{k-1})p(\boldsymbol{Y}_{k-1})}{p(\boldsymbol{y}_k|\boldsymbol{Y}_{k-1})p(\boldsymbol{Y}_{k-1})} \tag{7.17}$$

because knowledge of $\boldsymbol{Y}_{k-1}$ does not contribute to $\boldsymbol{y}_k$ if $\boldsymbol{x}_k$ is known. Cancelling the common factor in (7.17) gives

$$p(\boldsymbol{x}_k|\boldsymbol{Y}_k) = \frac{p(\boldsymbol{y}_k|\boldsymbol{x}_k)p(\boldsymbol{x}_k|\boldsymbol{Y}_{k-1})}{p(\boldsymbol{y}_k|\boldsymbol{Y}_{k-1})}. \tag{7.18}$$

Each of the three terms on the right hand side of this equation will be evaluated separately.

The p.d.f. $p(\boldsymbol{y}_k|\boldsymbol{x}_k)$ will be gaussian because $\boldsymbol{y}_k$ is equal to the sum of a zero-mean gaussian random variable $\boldsymbol{v}_k$ and a constant term $\boldsymbol{H}_o(k)\boldsymbol{x}_k$ (constant because $\boldsymbol{x}_k$ is given).

The mean of $p(\boldsymbol{y}_k|\boldsymbol{x}_k)$ will be the constant term $\boldsymbol{H}_o(k)\boldsymbol{x}_k$ and the variance will be the variance of $\boldsymbol{v}$, i.e. $\boldsymbol{V}_v$. Because $p(\boldsymbol{y}_k|\boldsymbol{x}_k)$ is gaussian, the first two moments completely specify the distribution. More formally

$$\mathcal{E}\{\boldsymbol{y}_k|\boldsymbol{x}_k\} = \boldsymbol{H}_o(k)\boldsymbol{x}_k \tag{7.19}$$

and

$$\mathrm{var}\{\boldsymbol{y}_k|\boldsymbol{x}_k\} = \boldsymbol{V}_v \tag{7.20}$$

where the variance matrix of a random vector is denoted by

$$\mathrm{var}\{\boldsymbol{z}\} \equiv \mathcal{E}\{[\boldsymbol{z} - \mathcal{E}\{\boldsymbol{z}\}][\boldsymbol{z} - \mathcal{E}\{\boldsymbol{z}\}]^{\mathrm{T}}\}. \tag{7.21}$$

Consider $p(\boldsymbol{y}_k|\boldsymbol{Y}_{k-1})$. Using (7.3), this can be written as

$$p(\boldsymbol{y}_k|\boldsymbol{Y}_{k-1}) = p(\boldsymbol{H}_0(k)\boldsymbol{x}_k + \boldsymbol{v}_k|\boldsymbol{Y}_{k-1}). \tag{7.22}$$

Now let us assume temporarily that $\boldsymbol{x}_k$ conditioned on $\boldsymbol{Y}_{k-1}$ is gaussian. Then $p(\boldsymbol{y}_k|\boldsymbol{Y}_{k-1})$ is a gaussian p.d.f, as $\boldsymbol{y}_k$ will be the weighted sum of two gaussian distributions, $\boldsymbol{H}_0(k)\boldsymbol{x}_k$ and $\boldsymbol{v}_k$. The mean of $\boldsymbol{y}_k|\boldsymbol{Y}_{k-1}$ will be given by

$$\mathcal{E}\{\boldsymbol{y}_k|\boldsymbol{Y}_{k-1}\} = \boldsymbol{H}_0(k)\mathcal{E}\{\boldsymbol{x}_k|\boldsymbol{Y}_{k-1}\} + \mathcal{E}\{\boldsymbol{v}_k|\boldsymbol{Y}_{k-1}\}. \tag{7.23}$$

We have assumed that $\boldsymbol{v}_k$ is a zero-mean white-noise process so that

$$\mathcal{E}\{\boldsymbol{y}_k|\boldsymbol{Y}_{k-1}\} = \boldsymbol{H}_0(k)\boldsymbol{x}_{k|k-1}, \tag{7.24}$$

where

$$\hat{\boldsymbol{x}}_{k|k-1} \equiv \mathcal{E}\{\boldsymbol{x}_k|\boldsymbol{Y}_{k-1}\}. \tag{7.25}$$

The variance of $\boldsymbol{y}_k|\boldsymbol{Y}_{k-1}$ is

$$\mathrm{var}\{\boldsymbol{y}_k|\boldsymbol{Y}_{k-1}\} = \mathcal{E}\{[\boldsymbol{y}_k - \boldsymbol{H}_0(k)\hat{\boldsymbol{x}}_{k|k-1}][\boldsymbol{y}_k - \boldsymbol{H}_0(k)\hat{\boldsymbol{x}}_{k|k-1}]^{\mathrm{T}}\}, \tag{7.26}$$

so using (7.3) gives

$$\begin{aligned}\mathrm{var}\{\boldsymbol{y}_k|\boldsymbol{Y}_{k-1}\} &= \mathcal{E}\{[\boldsymbol{H}_0(k)\boldsymbol{x}_k + \boldsymbol{v}_k - \boldsymbol{H}_0(k)\hat{\boldsymbol{x}}_{k|k-1}] \\ &\quad \cdot[\boldsymbol{H}_0(k)\boldsymbol{x}_k + \boldsymbol{v}_k - \boldsymbol{H}_0(k)\hat{\boldsymbol{x}}_{k|k-1}]^{\mathrm{T}}\}\end{aligned} \tag{7.27}$$

or

$$\begin{aligned}\mathrm{var}\{\boldsymbol{y}_k|\boldsymbol{Y}_{k-1}\} &= \boldsymbol{H}_0(k)\mathcal{E}\{\boldsymbol{x}_k\boldsymbol{x}_k^{\mathrm{T}} - \boldsymbol{x}_k\hat{\boldsymbol{x}}_{k|k-1}^{\mathrm{T}} - \hat{\boldsymbol{x}}_{k|k-1}\boldsymbol{x}_k^{\mathrm{T}} \\ &\quad +\hat{\boldsymbol{x}}_{k|k-1}\hat{\boldsymbol{x}}_{k|k-1}^{\mathrm{T}}\}\boldsymbol{H}_0^{\mathrm{T}}(k) + \mathrm{var}\{\boldsymbol{v}_k\}.\end{aligned} \tag{7.28}$$

The cross terms involving $\boldsymbol{v}_k$ are zero because $\boldsymbol{v}_k$ is assumed to be zero-mean white noise. If we denote $\boldsymbol{x}_k - \hat{\boldsymbol{x}}_{k|k-1}$ to be $\tilde{\boldsymbol{x}}_{k|k-1}$ then, after grouping terms, the above equation becomes

$$\mathrm{var}\{\boldsymbol{y}_k|\boldsymbol{Y}_{k-1}\} = \boldsymbol{H}_0(k)\mathrm{var}\{\tilde{\boldsymbol{x}}_{k|k-1}\}\boldsymbol{H}_0^{\mathrm{T}}(k) + \boldsymbol{V}_v(k) \tag{7.29}$$

or

$$\mathrm{var}\{\boldsymbol{y}_k|\boldsymbol{Y}_{k-1}\} = \boldsymbol{H}_0(k)\boldsymbol{V}_{\tilde{x}}(k|k-1)\boldsymbol{H}_0^{\mathrm{T}}(k) + \boldsymbol{V}_v(k) \tag{7.30}$$

where $\boldsymbol{V}_{\tilde{x}}(k|k-1) = \mathrm{var}\{\tilde{\boldsymbol{x}}_{k|k-1}\}$.

In order to evaluate (7.18) it now remains to find a suitable expression for $p(\boldsymbol{x}_k|\boldsymbol{Y}_{k-1})$. We have assumed that $\boldsymbol{x}_k|\boldsymbol{Y}_{k-1}$ is gaussian, so we need only to derive the mean and variance.

By using the state equation (7.1) we have

$$\mathcal{E}\{\boldsymbol{x}_k|\boldsymbol{Y}_{k-1}\} = \mathcal{E}\{\boldsymbol{A}(k,k-1)\boldsymbol{x}_{k-1} + \boldsymbol{\Gamma}(k-1)\boldsymbol{w}_{k-1}|\boldsymbol{Y}_{k-1}\}, \tag{7.31}$$

$$\mathcal{E}\{\boldsymbol{x}_k|\boldsymbol{Y}_{k-1}\} = \boldsymbol{A}(k,k-1)\mathcal{E}\{\boldsymbol{x}_{k-1}|\boldsymbol{Y}_{k-1}\} + \boldsymbol{\Gamma}(k-1)\mathcal{E}\{\boldsymbol{w}_{k-1}|\boldsymbol{Y}_{k-1}\}. \tag{7.32}$$

Define

$$\hat{\boldsymbol{x}}_k = \mathcal{E}\{\boldsymbol{x}_k|\boldsymbol{Y}_k\} \tag{7.33}$$

and note that $\mathcal{E}\{\boldsymbol{w}_{k-1}|\boldsymbol{Y}_{k-1}\}$ is zero because the best filtered estimate of zero-mean white noise is zero. Using these relationships (7.32) becomes

$$\mathcal{E}\{\boldsymbol{x}_k|\boldsymbol{Y}_{k-1}\} \equiv \hat{\boldsymbol{x}}_{k|k-1} = \boldsymbol{A}(k|k-1)\hat{\boldsymbol{x}}_{k-1}. \tag{7.34}$$

Now let us evaluate the variance of $\boldsymbol{x}_k|\boldsymbol{Y}_{k-1}$. Using the state equation and (7.34) we have that

$$\begin{aligned}\mathrm{var}\{\boldsymbol{x}_k|\boldsymbol{Y}_{k-1}\} &= \mathcal{E}\{[\boldsymbol{A}(k,k-1)\boldsymbol{x}_{k-1} + \boldsymbol{\Gamma}(k-1)\boldsymbol{w}_{k-1} - \boldsymbol{A}(k,k-1)\hat{\boldsymbol{x}}_{k-1}] \quad (7.35)\\ &\quad \cdot[\boldsymbol{A}(k,k-1)\boldsymbol{x}_{k-1} + \boldsymbol{\Gamma}(k-1)\boldsymbol{w}_{k-1} - \boldsymbol{A}(k|k-1)\hat{\boldsymbol{x}}_{k-1}]^{\mathrm{T}}|\boldsymbol{Y}_{k-1}\}\end{aligned}$$

so that

$$\begin{aligned}\mathrm{var}\{\boldsymbol{x}_k|\boldsymbol{Y}_{k-1}\} &= \boldsymbol{A}(k,k-1)\mathcal{E}\{(\boldsymbol{x}_{k-1} - \hat{\boldsymbol{x}}_{k-1})(\boldsymbol{x}_{k-1} - \hat{\boldsymbol{x}}_{k-1})^{\mathrm{T}}|\boldsymbol{Y}_{k-1}\}\boldsymbol{A}^{\mathrm{T}}(k,k-1)\\ &\quad +\boldsymbol{\Gamma}(k-1)\mathcal{E}\{\boldsymbol{w}_{k-1}\boldsymbol{w}_{k-1}^{\mathrm{T}}|\boldsymbol{Y}_{k-1}\}\boldsymbol{\Gamma}^{\mathrm{T}}(k-1) \qquad (7.36)\end{aligned}$$

where the cross terms have been set to zero. If we define

$$\tilde{\boldsymbol{x}}_k = \boldsymbol{x}_k - \hat{\boldsymbol{x}}_k \tag{7.37}$$

and

$$\boldsymbol{V}_{\tilde{x}}(k) = \mathrm{var}\{\tilde{\boldsymbol{x}}_k\} \tag{7.38}$$

then our expression for the variance of $\boldsymbol{x}_k|\boldsymbol{Y}_{k-1}$ becomes

$$\mathrm{var}\{\boldsymbol{x}_k|\boldsymbol{Y}_{k-1}\} = \boldsymbol{A}(k,k-1)\boldsymbol{V}_{\tilde{x}}(k-1)\boldsymbol{A}^{\mathrm{T}}(k,k-1)+\boldsymbol{\Gamma}(k-1)\boldsymbol{V}_w(k-1)\boldsymbol{\Gamma}^{\mathrm{T}}(k-1). \tag{7.39}$$

We have now calculated the mean and variance for each of the probability distributions in (7.18). Remember that the general form of the p.d.f. for a gaussian variable $\boldsymbol{x}$ with mean of $\bar{\boldsymbol{x}}$ and variance $\boldsymbol{V}$ is given by

$$C\exp\left(-\frac{1}{2}((\boldsymbol{x}-\bar{\boldsymbol{x}})^{\mathrm{T}}\boldsymbol{V}(\boldsymbol{x}-\bar{\boldsymbol{x}}))\right) \tag{7.40}$$

where C is a normalisation constant. Accordingly by using this form and Equations (7.19),(7.20),(7.30),(7.24), (7.34) and (7.39), we find that (7.18) can be expressed as

$$\begin{aligned}p(\boldsymbol{x}_k|\boldsymbol{Y}_k) &= C_1\exp\left(-\frac{1}{2}\{[\boldsymbol{y}_k - \boldsymbol{H}_o(k)\boldsymbol{x}_k]^{\mathrm{T}}\boldsymbol{V}_v^{-1}(k)[\boldsymbol{y}_k - \boldsymbol{H}_o(k)\boldsymbol{x}_k] \right. \quad (7.41)\\ &\quad +[\boldsymbol{x}_k - \hat{\boldsymbol{x}}_{k|k-1}]^{\mathrm{T}}\boldsymbol{V}_{\tilde{x}}^{-1}(k|k-1)[\boldsymbol{x}_k - \hat{\boldsymbol{x}}_{k|k-1}]\\ &\quad -[\boldsymbol{y}_k - \boldsymbol{H}_o(k)\hat{\boldsymbol{x}}_{k|k-1}]^{\mathrm{T}}[\boldsymbol{H}_o(k)\boldsymbol{V}_{\tilde{x}}(k|k-1)\boldsymbol{H}_o^{\mathrm{T}}\\ &\quad \left. +\boldsymbol{V}_v(k)]^{-1}[\boldsymbol{y}_k - \boldsymbol{H}_o(k)\hat{\boldsymbol{x}}_{k|k-1}]\}\right) \qquad (7.42)\end{aligned}$$

where C_1 is a normalisation constant. Here we have defined

$$V_{\tilde{x}}(k|k-1) \equiv \mathrm{var}\{x_k|Y_{k-1}\}. \tag{7.43}$$

After some algebraic manipulation, this can be put in the form

$$p(x_k|Y_k) = C_1 \exp\left(-\frac{1}{2}[x_k - \hat{x}_k]^{\mathrm{T}} V_{\tilde{x}}(k)[x_k - \hat{x}_k]\right) \tag{7.44}$$

where

$$\hat{x}_k = \hat{x}_{k|k-1} + K(k)(y(k) - H_o(k)\hat{x}_{k|k-1}), \tag{7.45}$$

$$V_{\tilde{x}}(k) = [H_o^{\mathrm{T}}(k)V_v^{-1}(k)H_o(k) + V_{\tilde{x}}(k|k-1)], \tag{7.46}$$

and

$$K(k) = V_{\tilde{x}}(k|k-1)H_o^{\mathrm{T}}(k)[V_v(k) + H_o(k)V_{\tilde{x}}(k|k-1)H_o^{\mathrm{T}}(k)]^{-1}. \tag{7.47}$$

$K(k)$ is often referred to as the gain of the Kalman filter.

At this point we can establish the correctness of our assumption concerning the gaussian nature of x_k conditioned on Y_{k-1}. The form of (7.44) shows that if this assumption is true then x_k conditioned on Y_k is also gaussian, but using the state equation and the gaussian nature of w_k means that x_{k+1} conditioned on Y_k will be gaussian. Hence, by using the fact that x_0 has been assumed to be gaussian, a simple proof by induction can be concluded.

Equation (7.44) represents the *a posteriori* probability. An inspection of the equation indicates that it will be maximised when the exponent is zero, which will occur when $x_k = \hat{x}_k$. Accordingly the maximum *a posteriori* estimator will be given by (7.45). This estimator is the Kalman filter.

7.2 Kalman Algorithm

In order to make practical use of the estimator derived in the previous section it is necessary to express $V_{\tilde{x}}(k)$ in terms of $K(k)$ and $V_{\tilde{x}}(k|k-1)$. After some manipulation [100], the result is

$$V_{\tilde{x}}(k) = [I - K(k)H_o(k)]V_{\tilde{x}}(k|k-1). \tag{7.48}$$

We can now state the algorithm to be used to implement a Kalman filter.

1. Input the initial conditions $\hat{x}_0 = \hat{x}_{0|0} = \mathcal{E}\{x_0\}$ and $V_{\tilde{x}}(0,0) = V_{\tilde{x}}(0|0) = V_x(0)$.

2. Set $j = 1$.

3. Calculate the *a priori* error variance using

$$V_{\tilde{x}}(j+1|j) = A(j+1,j)V_{\tilde{x}}(j)A^{\mathrm{T}}(j+1,j) + \Gamma(j)V_w(j)\Gamma^{\mathrm{T}}(j). \tag{7.49}$$

4. Evaluate the filter gain using

$$K(j) = V_{\tilde{x}}(j|j-1)H_o^{\mathrm{T}}(j)[H_o(j)V_{\tilde{x}}(j|j-1)H_o^{\mathrm{T}}(j) + V_v(j)]^{-1}. \quad (7.50)$$

5. Calculate the new estimate of $\hat{x}_j$ using

$$\hat{x}_j = A(j,j-1)\hat{x}_{j-1} + K(j)[y_j - H_o(j)A(j|j-1)\hat{x}_{j-1}]. \quad (7.51)$$

6. Calculate the *a posteriori* error variance using

$$V_{\tilde{x}}(j) = [I - K(j)H_o(j)]V_{\tilde{x}}(j|j-1). \quad (7.52)$$

7. Set $j = j + 1$ and go to Step 3.

Set out below is an example of how this algorithm may be used.

Example 7.2 Suppose a two-dimensional positioning system makes measurements on an object which moves with constant velocity with some process noise that induces small changes in the velocity. This might be a model for a plane flying at a constant altitude and subject to wind turbulence. The state equation for this situation will be [106]

$$z_{k+1} = Az_k + \Gamma w_k, \quad (7.53)$$

where

$$z = \left(x_1, \frac{dx_1}{dt}, x_2, \frac{dx_2}{dt}\right)^{\mathrm{T}}, \quad (7.54)$$

$$A = \begin{pmatrix} 1 & T & 0 & 0 \\ 0 & 1 & 0 & 0 \\ 0 & 0 & 1 & T \\ 0 & 0 & 0 & 1 \end{pmatrix}, \quad (7.55)$$

$$\Gamma = \begin{pmatrix} \frac{T}{2} & 0 \\ 1 & 0 \\ 0 & \frac{T}{2} \\ 0 & 1 \end{pmatrix}, \quad (7.56)$$

w is zero-mean gaussian white noise with covariance matrix V_w, T is the time interval between measurements, and x_1, x_2 are the cartesian coordinates of the object. The measurement model is

$$y_k = H_o x_k + v_k \quad (7.57)$$

where

$$H_0 = \begin{pmatrix} 1 & 0 & 0 & 0 \\ 0 & 0 & 1 & 0 \end{pmatrix} \quad (7.58)$$

(i.e. only spatial position measurements were available) and v is zero-mean gaussian noise with covariance matrix V_v.

A simulated trajectory was produced by inputting the following parameters to a computer model of the object's kinematics:

$$x_0 = (0, 2, 5, 3), \tag{7.59}$$

$$\boldsymbol{V}_w = \begin{pmatrix} 0.0025 & 0 \\ 0 & 0.0025 \end{pmatrix}, \tag{7.60}$$

and

$$\boldsymbol{V}_v = \begin{pmatrix} 25 & 0 \\ 0 & 25 \end{pmatrix}. \tag{7.61}$$

The simulated trajectory was then processed using the above Kalman filter algorithm. The initial parameters were

$$\hat{\boldsymbol{x}}_0 = (0, 0, 0, 0)^{\mathrm{T}} \tag{7.62}$$

and

$$\boldsymbol{V}_x(0) = \begin{pmatrix} 5 & 0 & 0 & 0 \\ 0 & 1 & 0 & 0 \\ 0 & 0 & 5 & 0 \\ 0 & 0 & 0 & 1 \end{pmatrix}. \tag{7.63}$$

A comparison of the actual, measured and estimated trajectories for x_1 (x-axis) and x_2 (y-axis) are shown in Fig. 7.1 and Fig. 7.2. It can be seen that when the kinematics is accurately known the Kalman filter gives excellent results.

If the object kinematics are not known then the results are not as good. Figure 7.3 shows the effect on the Kalman filter of unexpected kinematics. The trajectory followed by the object in this figure is the same as in Fig. 7.1 except that at $t = 50$ seconds the object makes a sharp turn, then resumes a constant velocity. It can be seen that during the turn the Kalman filter diverges severely from the actual trajectory. In this case the unfiltered data gives a better indication of position.

□

The Kalman filtering algorithm described here can be augmented in a number of ways. It is possible to correct for the effects of a deterministic driving force, a deterministic bias to the measurement model and correlation between $\boldsymbol{v}$ and $\boldsymbol{w}$ [100]. As well the analysis can be extended from the discrete to the continuous case [50].

One of the difficulties with a Kalman filter is that it requires a good model of the kinematics of the object. If the model is in error the filter can give very wrong results, as can be seen from Fig. 7.3 in Example 7.2. Models of ships, planes and cars are most likely to be wrong when the human pilot is undertaking course corrections (i.e. when the object is manoeuvering). A number of approaches have been proposed to overcome this problem [106, 4, 92]. In cases where there is little formal knowledge concerning the object kinematics, it is possible that a recurrent neural net [90] approach may be effective [52].

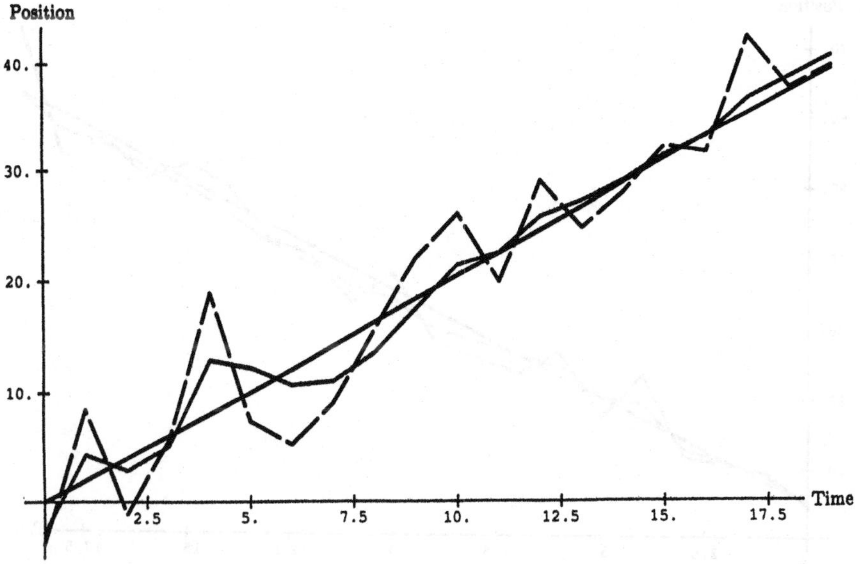

Fig. 7.1. A comparison of actual, estimated and measured trajectories for the x-axis. The solid dark line is actual, the solid light line is estimated and the dashed line is the measured data.

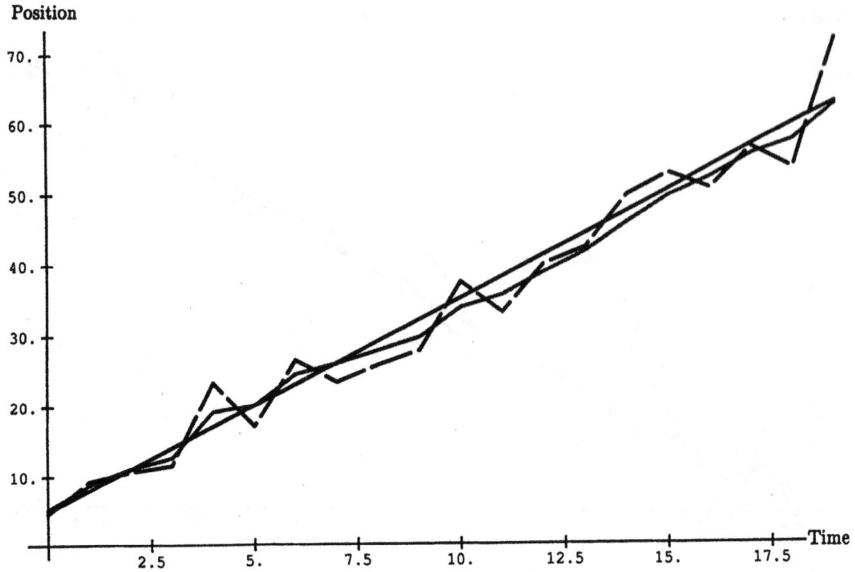

Fig. 7.2. A comparison of actual, estimated and measured trajectories for the y-axis. The solid dark line is actual, the solid light line is estimated and the dashed line is the measured data.

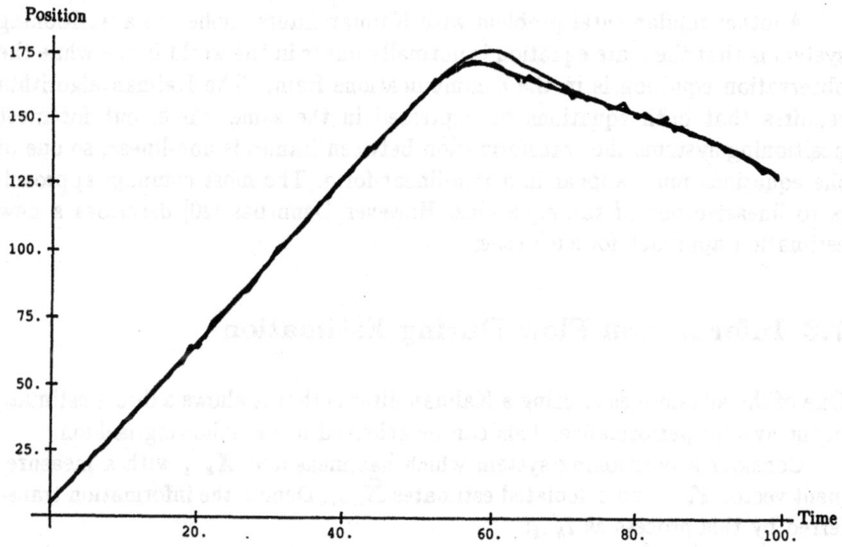

Fig. 7.3. A comparison of actual, estimated and measured trajectories for the x-axis. The solid dark line is actual, the solid light line is estimated and the dashed line is the measured data. There is a sharp turn at $t = 50$ s

Another fundamental problem with Kalman filters applied to a positioning system is that the state equation is normally linear in the world frame while the observation equation is in the communications frame. The Kalman algorithm requires that both equations be expressed in the same frame, but for most positioning systems the transformation between frames is non-linear, so one of the equations must appear in a non-linear form. The most common approach is to linearize one of the equations. However Demirbas [20] describes a new estimation approach for such cases.

7.3 Information Flow During Estimation

One of the advantages of using a Kalman filter is that it allows a direct estimate of the system performance. This can be achieved in the following fashion.

Consider a positioning system which has measured $\boldsymbol{X}_{k-1}$ with a measurement vector $\boldsymbol{Y}_{k-1}$ and calculated estimates $\widehat{\boldsymbol{X}}_{k-1}$. Denote the information transferred by this process as I_{k-1}:

$$I_{k-1} = I(\boldsymbol{X}_{k-1}; \widehat{\boldsymbol{X}}_{k-1}). \tag{7.64}$$

The question considered here is, given $\boldsymbol{Y}_{k-1}$ and a knowledge of the vehicle dynamics, how much additional information is sent by the reception of $\hat{\boldsymbol{x}}_k$. Denote this extra information by i_k. Formally:

$$i_k = I_k - I_{k-1}. \tag{7.65}$$

Now i_k can be calculated directly from the filter, in the case when $\boldsymbol{w}$, $\boldsymbol{v}$ and $\boldsymbol{x}_0$ have gaussian distributions and the system model and measurement mode are linear.

The process of sending the information for the kth estimate is as follows: Immediately prior to the kth estimate $\boldsymbol{Y}_{k-1}$ is known. The message $\boldsymbol{x}_k$ is transmitted, which is received as $\boldsymbol{y}_k$ and estimated to be $\hat{\boldsymbol{x}}_k$. Accordingly

$$i_k = I(\boldsymbol{x}_k; \hat{\boldsymbol{x}}_k | \boldsymbol{Y}_{k-1}) \tag{7.66}$$

where $I(\boldsymbol{x}_k; \hat{\boldsymbol{x}}_k | \boldsymbol{Y}_{k-1})$ is the conditional mutual information between $\boldsymbol{x}_k$ and $\widehat{\boldsymbol{x}_k}$, given $\boldsymbol{Y}_{k-1}$. Because $\hat{\boldsymbol{x}}_k = \boldsymbol{x}_k + \tilde{\boldsymbol{x}}$, we have that if $\hat{\boldsymbol{x}}_k$ and $\tilde{\boldsymbol{x}}_k$ are independent then $i_k = H(\boldsymbol{x}_k | \boldsymbol{Y}_{k-1}) - H(\tilde{\boldsymbol{x}}_k)$.

We will now prove that $\tilde{\boldsymbol{x}}_k$ and $\hat{\boldsymbol{x}}_k$ are independent. In Sect. 3.3 it was shown that an optimal entropy-error estimator will cause these two variables to be independent. Hence to prove they are independent in the case of the Kalman filter it is only necessary to prove that the optimal entropy-error estimator will be the same as the Kalman filter estimator.

This can be done in the following fashion. We have derived the Kalman estimator by using the maximum *a posteriori* method. Sage and Melsa [100, section 6.2] show that any estimator will provide the same result as the maximum a priori estimator [1] provided the cost function satisfies the following conditions:

[1] Sage and Melsa state the requirement for the minimum variance estimator but show it is gives the same result at the maximum a priori estimator

- the cost function is symmetric about $\tilde{x} = 0$,
- the cost function is convex,
- the *a posteriori* probability density function is symmetric about the maximum a priori estimate, and
- the product of the cost function and the a posteriori density goes to zero as $\tilde{x}$ goes to infinity.

Thus if we can show the entropy-error cost function satisfies the above conditions, then we know that the entropy-error estimator will be the same as the maximum *a posteriori* estimator and so $\widetilde{x_k}$ and $\widehat{x_k}$ will be independent.

The entropy-error cost function for this case is

$$C(\widetilde{x_k}) = -\log(p(\widetilde{x_k})). \tag{7.67}$$

First let us consider $\widetilde{x_k}|Y_k$. By a simple change of variable

$$p_{\widetilde{x}_k|Y_k}(\widetilde{x_k}|Y_k) = p_{x_k|Y_k}(\tilde{x}_k + \hat{x}_k|Y_k). \tag{7.68}$$

But from (7.44) we know that $x_k|Y_k$ is gaussian distributed with mean $\widehat{x_k}$ and covariance matrix $V_{\tilde{x}}(k)$. Accordingly we know that $p_{\widetilde{x}_k|Y_k}(\widetilde{x_k}|Y_k)$ will be a gaussian p.d.f with zero-mean and variance matrix $V_{\tilde{x}}(k)$. A check of the Kalman filter algorithm shows that $V_{\tilde{x}}(k)$ is independent of Y_k, so that $\tilde{x}_k$ will be gaussian distributed with zero mean and covariance matrix

$$\mathcal{E}\{\widetilde{x_k}\widetilde{x_k}^{\mathrm{T}}\} = V_{\tilde{x}_k}(k). \tag{7.69}$$

Such a gaussian density function will be clearly symmetric about $\widetilde{x_k} = 0$ and the negative logarithm will be convex. From (7.44) we know that the *a posteriori* p.d.f. is symmetric about $\hat{x}$, the maximum a posteriori estimate. Finally both $p_{\tilde{x}}(\tilde{x})$ and the *a posteriori* p.d.f. are gaussian with finite means and variances, so their product will go to zero as $\tilde{x}$ goes to infinity. Hence all four conditions are satisfied and we can conclude that the maximum *a posteriori* estimator will be the same as the error-entropy estimator. Hence $\widetilde{x_k}$ and $\widehat{x_k}$ will be independent and we can write (7.66) as

$$i_k = H(x_k|Y_{k-1}) - H(\widetilde{x_k}). \tag{7.70}$$

From our previous development, we know that $x_k|Y_{k-1}$ is gaussian distributed with covariance matrix $V_{\tilde{x}}(k|k-1)$ (see (7.39)). As well, x_k is gaussian distributed with covariance matrix $V_{\tilde{x}}(k)$ (see (7.69)).

The entropy of a n-variable gaussian distribution with covariance matrix A is given by (see Jones [53, page 151])

$$H = \frac{1}{2}\log((2\pi e)^n \det(A)) \tag{7.71}$$

so that

$$i_k = \frac{1}{2}\log\left[\frac{\det(V_{\tilde{x}}(k|k-1))}{\det(V_{\tilde{x}}(k))}\right], \tag{7.72}$$

or using (7.48)

$$i_k = -\frac{1}{2}\log(\det[\boldsymbol{I} - \boldsymbol{K}(k)\boldsymbol{H}_o(k)]). \tag{7.73}$$

Example 7.3 The purpose of this example is to calculate i_k for a simple positioning system. Suppose that in a one-dimensional positioning system, an object has the state equation

$$x_{k+1} = x_k + w, \tag{7.74}$$

where w is a zero-mean white gaussian process with variance σ_w^2, and x_0 is zero-mean gaussian distributed with variance σ_x^2. The measurement equation is

$$y_k = x_k + v, \tag{7.75}$$

where v is a zero-mean white gaussian process with variance σ_v^2.

Applying the Kalman filter algorithm described in Sect. 7.2 we have

$$V_{\tilde{x}}(0) = \sigma_x^2, \tag{7.76}$$

$$\hat{x}(0) = 0, \tag{7.77}$$

$$K(j) = \frac{V_{\tilde{x}}(j|j-1)}{(V_{\tilde{x}}(j|j-1) + \sigma_v^2)}, \tag{7.78}$$

$$V_{\tilde{x}}(j+1|j) = V_{\tilde{x}}(j) + \sigma_w^2, \tag{7.79}$$

and

$$V_{\tilde{x}}(j) = (1 - K(j))V_{\tilde{x}}(j|j-1). \tag{7.80}$$

Using these equations it is possible to evaluate i_j. Substituting (7.78) into (7.73) gives

$$i_j = -\frac{1}{2}\log\left[\frac{\sigma_v^2}{V_{\tilde{x}}(j|j-1) + \sigma_v^2}\right], \tag{7.81}$$

and by using of (7.79) this becomes

$$i_j = -\frac{1}{2}\log\left[\frac{\sigma_v^2}{V_{\tilde{x}}(j-1) + \sigma_w^2 + \sigma_v^2}\right], \tag{7.82}$$

or

$$i_j = \frac{1}{2}\log\left[1 + \frac{\sigma_w^2}{\sigma_v^2} + \frac{V_{\tilde{x}}(j-1)}{\sigma_v^2}\right]. \tag{7.83}$$

Let us find the difference equation governing $V_{\tilde{x}}(j-1)$, by substituting (7.79) into (7.80) and (7.78),

$$V_{\tilde{x}}(j) = (1 - K(j))(V_{\tilde{x}}(j-1) + \sigma_w^2), \tag{7.84}$$

$$K(j) = \frac{V_{\tilde{x}}(j-1) + \sigma_w^2}{(V_{\tilde{x}}(j-1) + \sigma_w^2 + \sigma_v^2)}. \tag{7.85}$$

Use (7.85) to eliminate $K(j)$ from (7.84):

$$V_{\tilde{x}}(j) = \frac{\sigma_v^2(V_{\tilde{x}}(j-1) + \sigma_w^2)}{V_{\tilde{x}}(j-1) + \sigma_w^2 + \sigma_v^2}. \tag{7.86}$$

This equation can be solved iteratively, using the initial condition $V_{\tilde{x}}(0) = \sigma_x^2$ and the result substituted into (7.83).

Consider the asymptotic form of (7.86). This equation will reach steady state when

$$V_{\tilde{x}}(j) - V_{\tilde{x}}(j-1) = 0, \tag{7.87}$$

so by substituting (7.86) into this equation we obtain the following quadratic equation

$$V_{\tilde{x}}(j)^2 + \sigma_w^2 V_{\tilde{x}}(j) - \sigma_v^2 + \sigma_w^2. \tag{7.88}$$

There are two possible solutions, but one can be ruled out as always being negative, accordingly

$$V_{\tilde{x}}(\infty) = \frac{-\sigma_w^2 + \sqrt{4\sigma_v^2\sigma_w^2 + \sigma_w^4}}{2} \tag{7.89}$$

where $V_{\tilde{x}}(\infty)$ denotes the asymptotic form.

It is of interest to examine $V_{\tilde{x}}(\infty)$ for the two cases, $\sigma_v^2 >> \sigma_w^2$ and $\sigma_v^2 << \sigma_w^2$. The result is as follows:

$$V_{\tilde{x}}(\infty) = \sigma_v\sigma_w \quad \text{if } \sigma_v^2 >> \sigma_w^2 \tag{7.90}$$

$$\sigma_v^2 \quad \text{if } \sigma_v^2 << \sigma_w^2, \tag{7.91}$$

so that substituting this equation into (7.83) gives

$$i_k = \frac{1}{2}\log\left(1 + \frac{\sigma_w^2}{\sigma_v^2} + \frac{\sigma_w}{\sigma_v}\right) \text{ if } \sigma_v^2 >> \sigma_w^2, \tag{7.92}$$

and

$$i_k = \frac{1}{2}\log\left[2 + \frac{\sigma_w^2}{\sigma_v^2}\right] \text{ if } \sigma_v^2 << \sigma_w^2. \tag{7.93}$$

In Appendix E there is a direct derivation of (7.93) which gives insight into the physical processes involved.

If $\sigma_v^2 << \sigma_w^2$ the asymptotic form is reached very quickly as can be seen from considering (7.86) when σ_w^2 is very large, the reason for this is that when σ_w^2 is large and σ_v^2 is small the object's kinematics tend to be governed by the stochastic element so there is little 'learning' needed by the filter.

The behaviour when $\sigma_v^2 >> \sigma_w^2$ can be understood by considering (7.86) when σ_w^2 is zero:

$$V_{\tilde{x}}(j) = \frac{\sigma_v^2(V_{\tilde{x}}(j-1))}{V_{\tilde{x}}(j-1) + \sigma_v^2}. \tag{7.94}$$

It is easy to show by induction that in this case

$$V_{\tilde{x}}(j) = \left(\frac{j}{\sigma_v^2} + \frac{1}{\sigma_x^2} \right)^{-1}. \tag{7.95}$$

It can be seen that this will approach the asymptotic form much more slowly, of the order of $\frac{1}{j}$. Here the objects kinematics are determined much more by the deterministic part, which is measured precisely by reducing the measurement error by averaging the measurement noise, a typical $\frac{1}{n}$ process.

The behaviour predicted by this simple system can be seen in the more complex system described in Example 7.2. Shown below are plots of the calculated differential information transfer, i_k for the two extreme cases of $V_v >> V_w$ and $V_v << V_w$. Figure 7.4 has

$$\boldsymbol{V_w} = \begin{pmatrix} 0.0025 & 0 \\ 0 & 0.0025 \end{pmatrix}, \tag{7.96}$$

and

$$\boldsymbol{V_v} = \begin{pmatrix} 25 & 0 \\ 0 & 25 \end{pmatrix}. \tag{7.97}$$

The slow approach to the asymptote can be seen from this figure. In calculating Fig. 7.5, the $\boldsymbol{V_v}$ matrix is the same but the $\boldsymbol{V_w}$ matrix was set to

$$\boldsymbol{V_w} = \begin{pmatrix} 2500 & 0 \\ 0 & 2500 \end{pmatrix}. \tag{7.98}$$

The asymptotic form is achieved after only one iteration. Note that the differential information transfer is much greater in this case, due to the large *a priori* uncertainty caused by the highly stochastic nature of the kinematics. □

The equation derived here for the differential information transfer is valid only if the appropriate gaussian distributions are assumed. However it should be possible to widen the results demonstrated here to allow Kalman filters to provide reasonable estimates of the differential information transfer.

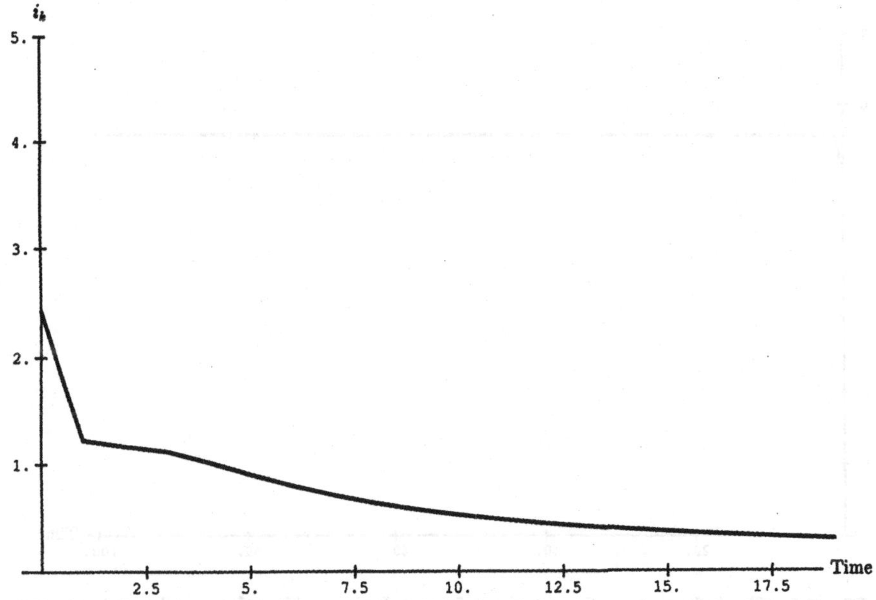

Fig. 7.4. The information flow derived from a Kalman filter for large measurement error.

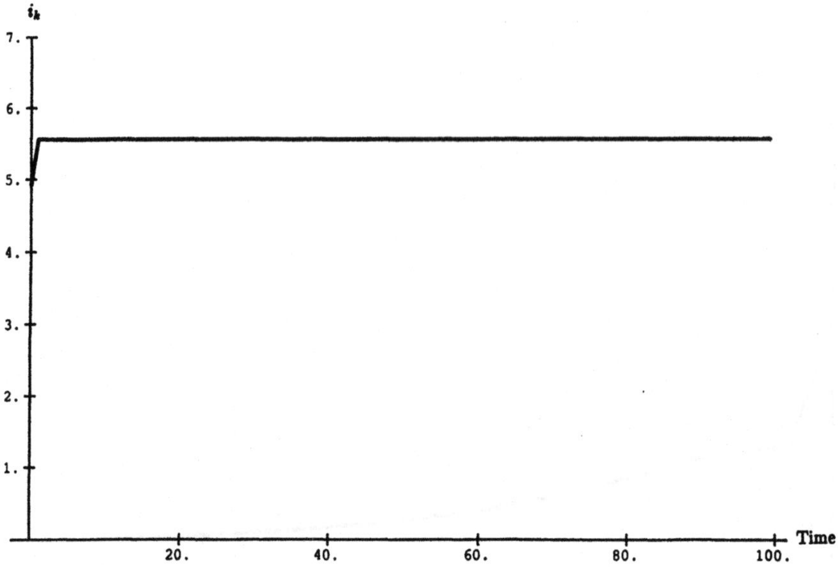

Fig. 7.5. The information flow derived from a Kalman filter for small measurement error.

8. Conclusion

This monograph has presented a method of analysing positioning systems. First a general formulation was derived and then this formulation was applied to wave-based systems (Chap. 2). This application to wave-based systems allowed a delineation of issues involved with source statistics, geometry, physical transmission channel and measurement strategy. A measure of performance (information performance) based on the average mutual information was presented. The chapter presented a notation that can be used to describe most forms of positioning system. If this notation were to be widely adopted it would facilitate the communications between researchers working on different forms of positioning systems.

Then in Chap. 3, an upper bound to the information performance was derived. This performance bound could be used by a system designer to establish the theoretical system capacity of a positioning device. In addition, in this chapter the invariance of information performance was established. The invariant nature of information performance considerably simplifies mathematical analysis of positioning systems. The most commonly used method of performance assessment (accuracy) is not invariant.

In Chap. 4 the formulation was used to determine optimal configuration of a positioning system, to compare two systems and to determine optimal measurement strategy. As well a technique for defining classes of systems was introduced. This latter development is particularly significant as it allows for the first time reasoning about classes of systems. This allows the development of general theorems about classes of systems and ultimately to the development of novel forms of system.

The amount of information produced by moving objects was derived in Chap. 5 by using the concept of source information rate. By comparing the source information rate with the known system capacity, a designer should be able to ensure that a system is capable of performing to within its specification. This may have important application to aircraft guidance and air defence systems, where system overload can be life threatening. The chapter also presented a method of classifying various methods of physical coding.

The issues involved with the decoding of a waveform to measure position were discussed in Chap. 6. The conventional measures of signal ambiguity and accuracy were outlined and a new definition of noise ambiguity was defined. This new method is much simpler to implement than the older method and has a direct analogue to the acquisition phase of many systems. Accordingly the noise ambiguity, as defined here, may become a more important system parameter. As well, a method of defining the bit rate and the probability of error of a positioning system was presented. This should facilitate direct comparisons of positioning systems and communications systems. Licensing authorities such as the Federal Communications Commission or the Australian Department of Trade and Commerce should be interested in such comparisons. Also this chapter derived a measure of coding performance called calibration performance which allows a practical way of comparing the coding efficiencies of various positioning systems. Next there was a discussion of a method of classifying performance enhancement schemes in terms of the characterisation advanced in this monograph. The chapter concluded by considering two further areas of investigation, waveform coding and realistic channels.

In Chap. 7 the process of estimation was placed in the context of this monograph and a method of calculating information performance derived from a Kalman filter. This allows performance to be calculated using the conventional tools of a positioning systems engineer, rather than needing to apply an information theoretic analysis.

In summary, a unified approach to the characterisation positioning systems has been formulated. This theory has satisfied all the requirement outlined in Chap. 1. It can be used to optimise and understand many aspects of system performance and so will be used in many different ways. The general formulation applies to any positioning system, and all of the results derived here are relevant to the most important type of positioning systems, wave-based positioning systems. Accordingly it has a wide field of applicability.

In order to emphasise the usefulness of the method of characterisation developed here, an example of how the method could be applied to a large and complex system is presented below.

Example 8.1 Suppose a company wishes to install a vehicle tracking system in a medium sized city. The system could have three applications: taxi dispatch, vehicle monitoring and emergency alarms. The operation of each of these is explained below:

Taxi Dispatch When a customer has called the taxi dispatch centre for a taxi, the customer's location is broadcast over the taxi radio network. Taxis which are unoccupied and near the customer will bid by pressing a button. This bid is sent over the normal radio network. The automatic dispatch system collects the first five bids and sends them to the vehicle tracking system. The vehicle tracking system measures the location of the five vehicles and works out which is closest to the customer. This information is sent back to the dispatch system and the closest taxi is dispatched. The expected number of dispatches per minute will be specified.

Vehicle Monitoring This requirement is to accurately and continuously track a number of vehicles. This function could be used by the police for surveillance work. The system must provide the facility to perform this task at different time intervals and accuracies.

Emergency Alarm This requirement is to allow the vehicle to signal via the vehicle tracking system that there is an emergency. For example a taxi driver might start the alarm if threatened by a customer or a anti-theft device could start the alarm if a car is stolen. This requirement means that the vehicle tracking system would not only have to send position information but an identification number so it is known which vehicle is being attacked/stolen.

The above description gives some idea of the complexity of modern tracking systems. The system has to perform a number of different functions, cover a wide geographic area, provide both location and conventional communication and a commercial system needs to do this at a low cost, within rigid bandwidth constraints.

This example will now work through the steps necessary to specify, design, implement and operate the system. To simplify the example, requirements that are not central to the arguments in this monograph are ignored.

The first step would be to collect statistics on the positional p.d.f. of the taxis, the vehicles likely to cause alarms and the vehicles to be continuously monitored. This could be done using historical data. Note that the positional p.d.f. will be highly non-uniform. For example in Sydney, Australia, it is very unlikely for a vehicle to drive on Sydney Harbour.

Next it is necessary to calculate the source information rate (see Sect. 5.1) of the three classes. In the case of the taxis and the alarms this is simple as each measurement of a vehicle should be approximately independent. In the case of the continuously monitored vehicles it will be necessary to make some experimental measurements of the spectra and use some of the bounding theorems described in Sect. 5.1. The designer will have to account for the effect of the customer requesting different accuracies.

After this it is necessary to work out the system requirements in terms of information performance. This would involve a modelling study of the cost benefit trade-offs for different levels of accuracy and information rate. The result should be a requirement that for the taxi system the information performance should be so many bits/s and an upper limit to accuracy (e.g 10 bits/s and the accuracy must be better than 100 metres). The information performance and accuracy need to be derived for the other two requirements. The conventional approach would have to specify a higher accuracy than 100 meters, because the designer is not given the additional specification in terms of bit rate. The specification is likely to include a necessary geographic coverage (e.g. 94% of the city's roadways must be covered by the tracking system).

The design team at this stage will consider a number of possible alternative technologies for example a spread spectrum radial-radial-radial system or a spread spectrum hyperbolic-hyperbolic-hyperbolic system. Both these would involve locating numerous reception sites around the city. The radial-radial-

radial system would measure the transit time to the vehicle from three different locations and calculate the intersection of the three circles (see Fig. 3.3). The hyperbolic-hyperbolic-hyperbolic system would measure the time difference of arrival at three independent pairs of sites. The independence requirement means that four sites are needed for an unambiguous fix. Hence the radial system would need less sites than the hyperbolic system, but the radial system has to measure round trip times so the in vehicle device has to be more expensive.

For each system being considered the team would consider the signal formats and signal waveforms to be used. To form an initial ball park estimate of the required spectrum the upper bound to system capacity (Sect. 3.2) could be calculated. As there is no velocity information to be monitored the selection of waveform is fairly straightforward, however if there was to be simultaneous velocity and position measurements then the method described in Sect. 6.8.2 could be developed for use.

Once the waveforms and signal formats are known the approximate calibration performance (Sect. 6.6) of each type of system can be calculated. This will give a broad indicator of performance and can be compared with the source information rate requirements. The bandwidth, power levels and other parameters will then be adjusted to make sure that source information rate can be handled. As well the calibration performance should be compared with the upper bound to system capacity. If there is a very large differential it may turn out that too much continuous monitoring (with long integration times to increase accuracy) is being requested.

The designers are now able to calculate the accuracy matrix $\boldsymbol{C}$ (see Sect. 6.3) for each system being considered. This matrix should be calculated as a function of signal strength to allow signal propagation data to be used.

For each of the systems, using the methods described in Chap. 4 and the a prior p.d.fs, it should be possible to work out the optimal location of reception sites and to compare the systems. As well the estimated information performance and accuracy will be compared with the specifications and appropriate adjustments made. At this point in time it should be possible to do cost benefit study and select the best system to use. Suppose that the decision is to use the radial system.

Another important calculation would be the system load due to the sending of alarms. There will be a trade-off between the position information and the alarm identification information. This trade-off will be easier to analyse than using conventional techniques because both are specified in bits/s.

At this point the designers could approach the licensing authorities to apply for a spectrum allocation. Their job will be much easier as the proposal to the licensing authority can be couched in terms of bit/s and so the spectral efficiency compared directly with conventional communications or other positioning systems which are vying for the same spectrum. This is not possible if the system is specified in terms of measurements per second of a particular accuracy. Indeed it is difficult to compare two different positioning systems if they have different accuracies and measurement rates without using the approach advocated in this monograph. Note that the licensing authorities are more likely to be concerned with the optimal calibration performance than the operational information performance. The former measures the capacity of the system, the latter how effectively that capacity is used.

Having obtained the promise of a license the designers would then start the detailed system design (assuming necessary funding!). The equipment would be designed, including the receivers. The concept of noise ambiguity (see Sect. 6.4) would be one of the parameters considered in this design. The bit rate and probability of error analyses (appropriately modified for real channels) could be used to establish the expected acquisition times for the system. For the continuous monitoring requirement, a Kalman filter may be used and the information rate from the Kalman filter calculated (see Sect. 7.3).

The designers would be likely to do a more detailed analysis of site location at this point. The methods of Chap. 4 would still form the basis of the analysis, but other factors such as coverage would also be considered. The designers will also calculate the system latencies (Chap. 2) and many other parameters and ensure they meet the constraints.

Having completed the design, the system will be constructed and installed. The system would be designed to output a continuous estimate of the system information performance. This would be used by operators as an overall indicator of system performance. Any major system failure will cause a sudden drop in this figure. As well the information performance will be monitored by the system owner as it will be a gross measure of system health and usage.

The forgoing is a highly idealised account of the design and operation of a large vehicle tracking system. However it does give some indication of the complexity of such systems and how the method of characterizing systems developed in this book can aid the designers, operators and licensing authorities to cope with that complexity. □

In the introduction to this monograph, the author stated that the work presented here should form the firm basis for a theory of positioning systems. Accordingly it is appropriate to conclude with a brief mention of the various areas of research that could be developed further.

- Overall, the work could benefit from a more general and rigourous application of Information Theory (Shannon Theory) and differential geometry.

- The wave-based formulation (Sect. 2.2) could be extended to account for the tracking of objects moving at relativistic velocities.

- The work on characterising classes of devices (Sect. 4.5) could be extended to encompass many other types of positioning systems.

- The use of information performance to optimise site location (Sect. 4.1) could be extended to include issues such as constraints on minimum accuracy and overall coverage.

- A rigourous proof of the multi-dimensional channel capacity could be derived (Sect. 3.2).
- Chaos seems a fruitful area for further research into the source information rate of moving objects (Chap. 5).
- When demand for urban vehicle positioning increases, the reliable multiplexing of multiple vehicles will become a greater issue as users seek to make maximum utilisation of available capacity (Sect. 3).
- There seems to be the promise of considerable performance enhancement for systems with low source information rates (Sect. 6.7).
- For application to urban vehicle tracking there is a need to extend the analysis to account for more realistic channels (Sect. 6.8).
- It should be possible to derive a tighter bound on the upper limit to the capacity of a channel, where the signal strength varies as a function of distance.
- Considerably more work could be done on deriving the capacity limits of multi-dimensional channels, in the cases where the simple linearisation used in Example 3.3 is not appropriate and/or the simple noise model is not applicable.
- The technique outlined in Sect. 6.8.2 may provide a useful method of wave form selection.
- There is a need to extend the overall analysis to account for the important area of multi-sensor fusion [96].

In conclusion, this monograph has presented a unified approach for the characterisation of positioning systems. It provides the basis for a theory of positioning systems and has identified many new areas of research.

A. Glossary of Commonly Used Symbols

$<\cdot>$	=	inner product
α	=	index to a class of positioning systems
A_c	=	accuracy in the communications frame
$A_g(R)$	=	ambiguity as a function of R
A_w	=	system accuracy in the world frame
β	=	r.m.s. bandwidth
β_n	=	generalised bandwith product
$\boldsymbol{\beta}(t)_j$	=	stochastic process representing the jth object
B	=	temporal bandwidth
b_e	=	bit error rate
B_i	=	spatial bandwidth in ith direction
b_r	=	bit rate
B_x	=	spatial bandwidth in x direction
B_y	=	spatial bandwith in y direction
$\mathcal{C}$	=	communications frame
χ	=	the squared modulus of χ is called the radar ambiguity function
C	=	maximum system capacity (bits per second)
$\delta(\cdot)$	=	Dirac delta function
$\boldsymbol{\delta x}$	=	small change in $\boldsymbol{x}$
$\mathcal{E}(\cdot)$	=	expectation
η	=	additional variable inserted to avoid ambiguity
e	=	dummy variable
E	=	total energy of signal
$\boldsymbol{f}(\boldsymbol{\phi})$	=	mapping from communications frame to world frame
$\boldsymbol{F}(\boldsymbol{Y})$	=	final estimator of position
γ_i	=	ordered triplet (t_i, l_i, j_i) which describes the ith measurement
$(\boldsymbol{g}_\alpha, \boldsymbol{q}_g, \boldsymbol{g}_g^{-1})$	=	the generator of the class
$\boldsymbol{g}(\boldsymbol{x})$	=	transformation from world frame to communications frame
$H(\boldsymbol{x})$	=	entropy of the random variable $\boldsymbol{x}$
$H(y\|x)$	=	the conditional entropy of y given x
$\boldsymbol{H}_0$	=	the matrix describing the measurement model
$I(\boldsymbol{x}; \boldsymbol{y})$	=	average mutual information between the random variables, $\boldsymbol{x}$ and $\boldsymbol{y}$
$i(t)$	=	instantaneous information performance
I_k	=	$I(\boldsymbol{X}_k, \widehat{\boldsymbol{X}_k})$

i_k	=	$I_k - I_{k-1}$
$\boldsymbol{J}_m(\boldsymbol{x})$	=	Jacobian matrix
J	=	absolute value of the determinant of the Jacobian matrix
j_i	=	the object being measured during the ith measurement
$\boldsymbol{K}(k)$	=	gain of Kalman filter
λ	=	wavelength
$\lambda_{i,K}$	=	ith eigenvalue of $\boldsymbol{\Phi}$
l_i	=	length of time of the ith measurement
L_i	=	dimension of view in the ith-direction
L_x	=	dimension of view in x direction
L_y	=	dimension of view in y direction
$\boldsymbol{M}_T$	=	The measurement vector
M	=	number of signals
$M(\frac{x}{\lambda})$	=	spatial response
$\boldsymbol{M}_\infty$	=	the measurement vector, M_T, when the number of measurements is infinity
M_x	=	average value of the logarithm of the Jacobian
n	=	noise power
N_0	=	one sided noise spectral density
N_g	=	noise power divided by the product of the bandwidths
$\boldsymbol{\Phi}_K$	=	vector of measured positions in communications frame
$\boldsymbol{\Phi}$	=	correlation matrix
ϕ	=	measurement of position in the communications frame
$p_{y\|x}(\boldsymbol{y}\|\boldsymbol{x})$	=	probability density function of $\boldsymbol{y}$ given $\boldsymbol{x}$
P_A	=	Woodward noise ambiguity
P_E	=	probability of error
P_T	=	information performance
$p_x(\boldsymbol{x})$	=	probability density function of $\boldsymbol{x}$
$Q_{\phi\|\xi}$	=	entropy of conditional error in communications frame
$q(\boldsymbol{\xi})$	=	random transformation representing measurement error in the communications frame
$\boldsymbol{R}(k)$	=	n by n sub-matrix of $\boldsymbol{\Phi}_K$
$\rho(\boldsymbol{x}, \boldsymbol{y})$	=	distortion measure between $\boldsymbol{x}$ and $\boldsymbol{y}$.
R	=	$\frac{2S_pT_g}{N_g}$
$R(D)$	=	rate distortion function
$\boldsymbol{S}$	=	class of positioning systems
σ	=	standard deviation
S_p	=	signal power
t	=	time
T	=	time period over which measurements are made
T_g	=	observation volume
t_i	=	time that the ith measurement is taken
Trc	=	trace of matrix
$u(\boldsymbol{w}, \boldsymbol{\xi})$	=	signal that is a function of position and a vector of parameters $\boldsymbol{w}$
$u_i^{'}$	=	partial derivative of $u(\boldsymbol{w}, \boldsymbol{\xi})$ with respect to ξ_i

$\boldsymbol{u}_i$	=	control parameters
$v(\boldsymbol{w})$	=	received signal
V_n	=	domain of interest in world frame
V_u	=	covariance matrix for measurement noise in Kalman filter analysis
V_w	=	covariance matrix for kinematical noise in Kalman filter analysis
$\mathcal{W}$	=	world frame
$\widehat{\boldsymbol{X}}$	=	vector of estimated positions
Ξ	=	vector of positions in communications frame
$\hat{\boldsymbol{x}}$	=	estimated position of object
$\boldsymbol{\xi}$	=	position in communications frame, prior to measurement error
$\boldsymbol{x}_j$	=	actual position of the jth measurement
$\boldsymbol{X}$	=	vector of positions
$\boldsymbol{y}$	=	'raw' (without estimation) measurement in world frame
$\boldsymbol{Y}$	=	vector of measured positions in world frame
$var(\boldsymbol{z})$	=	variance matrix of a random vector
z_i	=	output of ith correlator

B. Review of Information Theory

The following is a very brief recapitulation of some of the important definitions and theorems in elementary information theory. The recapitulation follows an excellent introduction to this subject which is given by Jones [53]. A more advanced treatment (and all the results presented here) may be found in Fano [29].

Consider a system, S, which consists of a set of events $E_1, \ldots, E_n$. The probability of the kth event is given by p_k. Each of the p_k will be greater than or equal to zero and

$$\sum_{k=1}^{n} p_k = 1. \tag{B.1}$$

The entropy of S is a measure of the uncertainty in the system. It is defined as

$$H(S) = -\sum_{k=1}^{n} p_k \log p_k. \tag{B.2}$$

By convention, if p_k is zero, then $p_k \log p_k$ is set to zero. Clearly, $H(S) \geq 0$.

Now consider two systems, S_1 and S_2. The events in S_1 are $E_1, \ldots, E_n$ with probabilities $p_1, \ldots, p_n$. Similarly, the events in S_2 are $F_1, \ldots, F_m$ with probabilities $q_1, \cdots, q_m$.

Define the connection between the two systems as

$$P(E_j \cap F_k) = p_{ij};\; j = 1, \ldots, n;\; k = 1, \ldots, m \tag{B.3}$$

with $p_{ij} \geq 0$ and

$$\sum_{j=1}^{n}\sum_{k=1}^{m} p_{jk} = 1. \tag{B.4}$$

The average mutual information between S_1 and S_2 is defined as

$$I(S_1; S_2) = \sum_{j=1}^{n}\sum_{k=1}^{m} p_{jk} \log\left(\frac{p_{jk}}{p_j p_k}\right). \tag{B.5}$$

For the case of a continuous random variable $\{x\}$ with probability density function $f_{\{x\}}$, the entropy is defined as

$$H(\{x\}) = -\int_{-\infty}^{\infty} dx\, f_{\{x\}}(x) \log f_{\{x\}}(x). \tag{B.6}$$

Note that the entropy of a continuous distribution need not be zero (because $f_{\{x\}}(x)$ can be greater than 1) and that the entropy is not necessarily invariant to a change in variables.

As discussed in the main part of this book, our notation becomes very cluttered due to the generality of the formulation. Therefore the notation has been simplified by omitting the $\{\cdot\}$ notation for random variable. We will follow this convention for the remainder of this Appendix.

The average mutual information between two continuous random variables, x and y is given by

$$I(x;y) = \int_{-\infty}^{\infty}\int_{-\infty}^{\infty} \mathrm{d}x \; \mathrm{d}y \, f_{x,y}(x,y) \log\left(\frac{f_{x,y}(x,y)}{f_x(x)f_y(y)}\right). \tag{B.7}$$

Because of the symmetry of this definition we see that $I(x,y) = I(y,x)$.

The quantity $H(x|y)$ is defined by

$$H(x|y) = \int_{-\infty}^{\infty}\int_{-\infty}^{\infty} \mathrm{d}x \; \mathrm{d}y \, f_{x,y}(x,y) \log\left(\frac{f_{x,y}(x,y)}{f_x(x)f_y(y)}\right) \tag{B.8}$$

and $H(x,y)$ is defined as

$$H(x,y) = \int_{-\infty}^{\infty}\int_{-\infty}^{\infty} \mathrm{d}x \; \mathrm{d}y \, f_{x,y}(x,y) \log(f_{x,y}(x,y)). \tag{B.9}$$

The following useful results will be used in the text:

Result 1 *The probability density function which gives the greatest entropy subject to restriction of finite variance, σ^2, is the gaussian distribution (with variance equal to σ^2) [53, Theorem 6.2].*

Result 2 *$I(x,y) \geq 0$ with equality if and only if the random variables x and y are statistically independent [53, Theorem 6.3].*

Result 3 *For two random variable x and y, we have that [53, p143]*

$$I(x,y) = H(y) - H(y|x). \tag{B.10}$$

Result 4 *If x,y and z are random variables, such that $z = x + y$ and x and y are independent, we have that [53, p149]*

$$I(x,z) = H(z) - H(y). \tag{B.11}$$

C. Proof of Proposition 3.5

$I(\Xi;\Phi)$ is given by

$$I(\Xi;\Phi) = \int_{U_n} \mathrm{d}\Xi \int_{U_n} \mathrm{d}\Phi \, p_c(\Xi,\Phi) \log\left[\frac{p_c(\Xi,\Phi)}{(p_\Xi(\Xi) p_\Phi(\Phi))}\right] \tag{C.1}$$

where $p_c(\Xi,\Phi)$ is the joint p.d.f. of Ξ and Φ and $p_\Xi(\Xi)$,$p_\Phi(\Phi)$ are the p.d.fs of Ξ and Φ respectively. Also U_n is the communications frame region corresponding to V_n.

Now generalising the method of Jones [53], gives

$$p(X,Y) = p_c(\Xi,\Phi) J(X,Y) \tag{C.2}$$

where $J(X,Y)$ is the Jacobian defined [1] as follows:

$$J(X,Y) = \left|\det\begin{pmatrix} G(X) & G(Y) \\ F^{-1}(X) & F^{-1}(Y) \end{pmatrix}\right| \tag{C.3}$$

where $G(X), G(Y), F^{-1}(X), F^{-1}(Y)$ are defined as follows:

$$\boldsymbol{G}(\boldsymbol{X}) = \begin{pmatrix} \boldsymbol{G}(\boldsymbol{x}_1;\boldsymbol{x}_1) & \dots & \boldsymbol{G}(\boldsymbol{x}_1,\boldsymbol{x}_K) \\ \vdots & & \vdots \\ \boldsymbol{G}(\boldsymbol{x}_K;\boldsymbol{x}_1) & \dots & \boldsymbol{G}(\boldsymbol{x}_K,\boldsymbol{x}_K) \end{pmatrix}. \tag{C.4}$$

where

$$\boldsymbol{G}(\boldsymbol{x}_i;\boldsymbol{x}_j) = \begin{pmatrix} \frac{\partial g_1(\boldsymbol{x}_i)}{\partial x_{1j}} & \dots & \frac{\partial g_1(\boldsymbol{x}_i)}{\partial x_{nj}} \\ \vdots & & \vdots \\ \frac{\partial g_n(\boldsymbol{x}_i)}{\partial x_{1j}} & \dots & \frac{\partial g_n(\boldsymbol{x}_i)}{\partial x_{nj}} \end{pmatrix}. \tag{C.5}$$

$$\boldsymbol{G}(\boldsymbol{Y}) = \begin{pmatrix} \boldsymbol{G}(\boldsymbol{x}_1;\boldsymbol{y}_1) & \dots & \boldsymbol{G}(\boldsymbol{x}_1,\boldsymbol{y}_K) \\ \vdots & & \vdots \\ \boldsymbol{G}(\boldsymbol{x}_K;\boldsymbol{y}_1) & \dots & \boldsymbol{G}(\boldsymbol{x}_K,\boldsymbol{y}_K) \end{pmatrix}, \tag{C.6}$$

where

$$\boldsymbol{G}(\boldsymbol{x}_i;\boldsymbol{y}_j) = \begin{pmatrix} \frac{\partial g_1(\boldsymbol{x}_i)}{\partial y_{1j}} & \dots & \frac{\partial g_1(\boldsymbol{x}_i)}{\partial y_{nj}} \\ \vdots & & \vdots \\ \frac{\partial g_n(\boldsymbol{x}_i)}{\partial y_{1j}} & \dots & \frac{\partial g_n(\boldsymbol{x}_i)}{\partial y_{nj}} \end{pmatrix}, \tag{C.7}$$

---

[1] For notational simplicity this differs slightly from the conventional Jacobian, in that the absolute value is taken.

$$\boldsymbol{F}^{-1}(\boldsymbol{X}) = \begin{pmatrix} \boldsymbol{F}^{-1}(\boldsymbol{x}_1;\boldsymbol{x}_1) & \dots & \boldsymbol{F}^{-1}(\boldsymbol{x}_1,\boldsymbol{x}_K) \\ \vdots & & \vdots \\ \boldsymbol{F}^{-1}(\boldsymbol{x}_K;\boldsymbol{x}_1) & \dots & \boldsymbol{F}^{-1}(\boldsymbol{x}_K,\boldsymbol{x}_K) \end{pmatrix}, \tag{C.8}$$

where

$$\boldsymbol{F}^{-1}(\boldsymbol{x}_i;\boldsymbol{x}_j) = \begin{pmatrix} \frac{\partial f_1^{-1}(\boldsymbol{x}_i)}{\partial \boldsymbol{x}_{1j}} & \dots & \frac{\partial f_1^{-1}(\boldsymbol{x}_i)}{\partial \boldsymbol{x}_{nj}} \\ \vdots & & \vdots \\ \frac{\partial f_n^{-1}(\boldsymbol{x}_i)}{\partial \boldsymbol{x}_{1j}} & \dots & \frac{\partial f_n^{-1}(\boldsymbol{x}_i)}{\partial \boldsymbol{x}_{nj}} \end{pmatrix}. \tag{C.9}$$

$$\boldsymbol{F}^{-1}(\boldsymbol{Y}) = \begin{pmatrix} \boldsymbol{F}^{-1}(\boldsymbol{x}_1;\boldsymbol{y}_1) & \dots & \boldsymbol{F}^{-1}(\boldsymbol{x}_1,\boldsymbol{y}_K) \\ \vdots & & \vdots \\ \boldsymbol{F}^{-1}(\boldsymbol{x}_K;\boldsymbol{y}_1) & \dots & \boldsymbol{F}^{-1}(\boldsymbol{x}_K,\boldsymbol{y}_K) \end{pmatrix}, \tag{C.10}$$

where

$$\boldsymbol{F}^{-1}(\boldsymbol{x}_i;\boldsymbol{y}_j) = \begin{pmatrix} \frac{\partial f_1^{-1}(\boldsymbol{x}_i)}{\partial \boldsymbol{y}_{1j}} & \dots & \frac{\partial f_1^{-1}(\boldsymbol{x}_i)}{\partial \boldsymbol{y}_{nj}} \\ \vdots & & \vdots \\ \frac{\partial f_n^{-1}(\boldsymbol{x}_i)}{\partial \boldsymbol{y}_{1j}} & \dots & \frac{\partial f_n^{-1}(\boldsymbol{x}_i)}{\partial \boldsymbol{y}_{nj}} \end{pmatrix}. \tag{C.11}$$

Given the assumption that the cross-derivatives are zero (see (3.35),(3.34)), this becomes

$$J(\boldsymbol{X},\boldsymbol{Y}) = \left|\det\begin{pmatrix} \boldsymbol{G}(\boldsymbol{X}) & 0 \\ 0 & \boldsymbol{F}^{-1}(\boldsymbol{Y}) \end{pmatrix}\right| \tag{C.12}$$

But

$$J(\boldsymbol{X},\boldsymbol{Y}) = \left|\det\left(\begin{pmatrix} \boldsymbol{G}(\boldsymbol{X}) & 0 \\ 0 & \boldsymbol{I} \end{pmatrix}\begin{pmatrix} \boldsymbol{I} & 0 \\ 0 & \boldsymbol{F}^{-1}(\boldsymbol{Y}) \end{pmatrix}\right)\right|, \tag{C.13}$$

where $\boldsymbol{I}$ is the identity matrix. Using the fact that the determinant of the product of two square matrices equals the product of the determinant of the individual matrices we have

$$J(\boldsymbol{X},\boldsymbol{Y}) = |\det(\boldsymbol{G}(\boldsymbol{X}))|\left|\det\left(\boldsymbol{F}^{-1}(\boldsymbol{Y})\right)\right|. \tag{C.14}$$

Defining $J(\boldsymbol{X}) = |\det(\boldsymbol{G}(\boldsymbol{X}))|$ and $J(\boldsymbol{Y}) = |\det(\boldsymbol{F}^{-1}(\boldsymbol{Y}))|$, gives the simple form

$$J(\boldsymbol{X},\boldsymbol{Y}) = J(\boldsymbol{X})J(\boldsymbol{Y}). \tag{C.15}$$

Now

$$p_{\Xi}(\boldsymbol{\Xi}) = \int_{U^n} p_c(\boldsymbol{\Xi},\boldsymbol{\Phi})\,\mathrm{d}\boldsymbol{\Phi}. \tag{C.16}$$

If we substitute (C.15) and (C.2) into (C.16), we have

$$p_{\Xi}(\boldsymbol{\Xi}) = \int_{U^n} \frac{p(\boldsymbol{X},\boldsymbol{Y})}{J(\boldsymbol{X})J(\boldsymbol{Y})}\,\mathrm{d}\boldsymbol{\Phi}. \tag{C.17}$$

A change of variable from $\boldsymbol{\Phi}$ to $\boldsymbol{Y}$ and integrating yields

$$p_{\Xi}(\boldsymbol{\Xi}) = \frac{p_X(\boldsymbol{X})}{J(\boldsymbol{X})}. \tag{C.18}$$

Similarly we find that

$$p_{\Phi}(\boldsymbol{\Phi}) = \frac{p_Y(\boldsymbol{Y})}{J(\boldsymbol{Y})} \tag{C.19}$$

Combining (C.2), (C.19) and (C.18) gives

$$\frac{p_c(\boldsymbol{\Xi}, \boldsymbol{\Phi})}{p_{\Xi}(\boldsymbol{\Xi}) p_{\Phi}(\boldsymbol{\Phi})} = \frac{p(\boldsymbol{X}, \boldsymbol{Y})}{p_X(\boldsymbol{X}) p_Y(\boldsymbol{Y})}. \tag{C.20}$$

Substituting (C.2), (C.15) and (C.20) into (C.1) gives

$$I(\boldsymbol{\Xi}; \boldsymbol{\Phi}) = \int_{U_n} \mathrm{d}\boldsymbol{\Xi} \int_{U_n} \mathrm{d}\boldsymbol{\Phi} \, \frac{p(\boldsymbol{X}, \boldsymbol{Y})}{J(\boldsymbol{X}) J(\boldsymbol{Y})} \log \left(\frac{p(\boldsymbol{X}, \boldsymbol{Y})}{p_X(\boldsymbol{X}) p_Y(\boldsymbol{Y})} \right). \tag{C.21}$$

Changing variables yields

$$I(\boldsymbol{\Xi}; \boldsymbol{\Phi}) = \int_{V_n} \mathrm{d}\boldsymbol{X} \int_{V_n} \mathrm{d}\boldsymbol{Y} \, p(\boldsymbol{X}, \boldsymbol{Y}) \log \left(\frac{p(\boldsymbol{X}, \boldsymbol{Y})}{p_X(\boldsymbol{X}) p_Y(\boldsymbol{Y})} \right). \tag{C.22}$$

A comparison with (2.6), shows that indeed

$$I(\boldsymbol{\Xi}; \boldsymbol{\Phi}) = I(\boldsymbol{X}; \boldsymbol{Y}). \tag{C.23}$$

This concludes the proof of the proposition.

D. Proof of Proposition 3.6

First it is necessary to prove that

$$I(\boldsymbol{X}';\boldsymbol{Y}') = I(\boldsymbol{\Xi}';\boldsymbol{\Phi}'). \tag{D.1}$$

In order to do this it is necessary to define the Jacobian of the augmented frames. For the image of the ith measurement with respect to the jth measurement[1], this is given by

$$\boldsymbol{J}'(\boldsymbol{x}_i;\boldsymbol{x}_j) = \mathcal{V}_n\left[\boldsymbol{A}(\boldsymbol{x}_i;\boldsymbol{x}_i)\right], \tag{D.2}$$

where

$$\boldsymbol{A}(\boldsymbol{x}_i;\boldsymbol{x}_j) = \begin{pmatrix} \frac{\partial g_1(\boldsymbol{x}_i)}{\partial x_{1j}} & \cdots & \frac{\partial g_1(\boldsymbol{x}_i)}{\partial x_{nj}} \\ \vdots & & \vdots \\ \frac{\partial g_{n+1}(\boldsymbol{x}_i)}{\partial x_{1j}} & \cdots & \frac{\partial g_{n+1}(\boldsymbol{x}_i)}{\partial x_{nj}} \end{pmatrix} \tag{D.3}$$

and $\mathcal{V}_n\left[\cdot\right]$ is the n-dimensional 'volume' of the volume element generated by the columns of the matrix.

The formula for calculating this volume will depend on the dimensionality. For example if $n = 1$, $\boldsymbol{A}$ will consist of a single two-dimensional column vector. The 'volume' will be the length of this vector. If $n = 2$, $\boldsymbol{A}$ will consist of two column vectors. Each vector will have three elements. These two vectors generate an element of surface area in the augmented communications frame. The 'volume' of this element is given by the magnitude of cross product of the two vectors [124].

In a similar fashion it is possible to define $\boldsymbol{A}(\boldsymbol{y}_i;\boldsymbol{y}_j)$ and $\boldsymbol{J}'(\boldsymbol{y}_i;\boldsymbol{y}_j)$ and so we have

$$\boldsymbol{J}'(\boldsymbol{X},\boldsymbol{Y}) = \mathcal{V}_n[\boldsymbol{A}(\boldsymbol{X},\boldsymbol{Y})] \tag{D.4}$$

where

$$\boldsymbol{A}(\boldsymbol{X},\boldsymbol{Y}) = \begin{pmatrix} \boldsymbol{A}(\boldsymbol{x}_1;\boldsymbol{x}_1) & \cdots & \boldsymbol{A}(\boldsymbol{x}_1;\boldsymbol{x}_K) & 0 & \cdots & 0 \\ \vdots & & \vdots & \vdots & & \vdots \\ \boldsymbol{A}(\boldsymbol{x}_K;\boldsymbol{x}_1) & \cdots & \boldsymbol{A}(\boldsymbol{x}_K;\boldsymbol{x}_K) & 0 & \cdots & 0 \\ 0 & \cdots & 0 & \boldsymbol{A}(\boldsymbol{y}_1;\boldsymbol{y}_1) & \cdots & \boldsymbol{A}(\boldsymbol{y}_1;\boldsymbol{y}_K) \\ \vdots & & \vdots & \vdots & & \vdots \\ 0 & \cdots & 0 & \boldsymbol{A}(\boldsymbol{y}_K;\boldsymbol{y}_1) & \cdots & \boldsymbol{A}(\boldsymbol{y}_K;\boldsymbol{y}_K) \end{pmatrix} \tag{D.5}$$

[1] This quantity will normally be zero unless $i = j$.

Now, we can write that

$$\mathcal{V}_n[A(X,Y)] = \mathcal{V}_n\left[\begin{pmatrix} A(\boldsymbol{x}_1;\boldsymbol{x}_1) & \dots & A(\boldsymbol{x}_1;\boldsymbol{x}_K) \\ \vdots & & \vdots \\ A(\boldsymbol{x}_K;\boldsymbol{x}_1) & \dots & A(\boldsymbol{x}_K;\boldsymbol{x}_K) \\ 0 & \dots & 0 \\ \vdots & & \vdots \\ 0 & \dots & 0 \end{pmatrix}\right] \cdot \mathcal{V}_n\left[\begin{pmatrix} 0 & \dots & 0 \\ \vdots & & \vdots \\ 0 & \dots & 0 \\ A(\boldsymbol{y}_1;\boldsymbol{y}_1) & \dots & A(\boldsymbol{y}_1;\boldsymbol{y}_K) \\ \vdots & & \vdots \\ A(\boldsymbol{y}_K;\boldsymbol{y}_1) & \dots & A(\boldsymbol{y}_K;\boldsymbol{y}_K) \end{pmatrix}\right]. \tag{D.6}$$

This follows because the two halves of the matrix are orthogonal.

Accordingly

$$J'(X,Y) = J'(X)J'(Y) \tag{D.7}$$

where

$$J'(X) = \mathcal{V}_n\left[\begin{pmatrix} A(\boldsymbol{x}_1;\boldsymbol{x}_1) & \dots & A(\boldsymbol{x}_1;\boldsymbol{x}_K) \\ \vdots & & \vdots \\ A(\boldsymbol{x}_K;\boldsymbol{x}_1) & \dots & A(\boldsymbol{x}_K;\boldsymbol{x}_K) \\ 0 & \dots & 0 \\ \vdots & & \vdots \\ 0 & \dots & 0 \end{pmatrix}\right] \tag{D.8}$$

and

$$J'(Y) = \mathcal{V}_n\left[\begin{pmatrix} 0 & \dots & 0 \\ \vdots & & \vdots \\ 0 & \dots & 0 \\ A(\boldsymbol{y}_1;\boldsymbol{y}_1) & \dots & A(\boldsymbol{y}_1;\boldsymbol{y}_K) \\ \vdots & & \vdots \\ A(\boldsymbol{y}_K;\boldsymbol{y}_1) & \dots & A(\boldsymbol{y}_K;\boldsymbol{y}_K) \end{pmatrix}\right] \tag{D.9}$$

Hence the Jacobian is separable, and the proof may run as in Proposition 3, so it is established that

$$I(X';Y') = I(\Xi';\Phi'). \tag{D.10}$$

Hence to prove the proposition, it is only necessary to establish that

$$I(X;Y) = I(X';Y'). \tag{D.11}$$

From (2.6), we have that

$$I(X';Y') = \int_{V_n'} \mathrm{d}\boldsymbol{x}_1' \dots \int_{V_n'} \mathrm{d}\boldsymbol{x}_K'\, p(X',Y') \log\left(\frac{p'(X',Y')}{p_X'(X')p_Y'(Y')}\right) \tag{D.12}$$

But we have that

$$\frac{p'(\boldsymbol{X}',\boldsymbol{Y}')}{p'_X(\boldsymbol{X}')p'_Y(\boldsymbol{Y}')} = \frac{p(\boldsymbol{X},\boldsymbol{Y})\delta(e_1)\dots\delta(e_K)\delta(f_1)\dots\delta(f_K)}{(p_X(\boldsymbol{X})\delta(e_1)\dots\delta(e_K))(p_Y(\boldsymbol{Y})\delta(f_1)\dots(f_K))} \tag{D.13}$$

so that

$$\frac{p'(\boldsymbol{X}',\boldsymbol{Y}')}{p'_X(\boldsymbol{X}')p'_Y(\boldsymbol{Y}')} = \frac{p(\boldsymbol{X},\boldsymbol{Y})}{p_X(\boldsymbol{X})p_Y(\boldsymbol{Y})} \tag{D.14}$$

Substituting this identity into (D.12) gives

$$I(\boldsymbol{X}';\boldsymbol{Y}') = \int_{V'_n} \mathrm{d}\boldsymbol{X}'_1 \int_{V'_n} \mathrm{d}\boldsymbol{Y}'_K\, p(\boldsymbol{X}',\boldsymbol{Y}_i) \log\left(\frac{p(\boldsymbol{X},\boldsymbol{Y})}{p_X(\boldsymbol{X})p_Y(\boldsymbol{Y})}\right) \tag{D.15}$$

or

$$\begin{aligned} I(\boldsymbol{X}';\boldsymbol{Y}') &= \int_{V'_n}\dots\int_{V'_n} \mathrm{d}\boldsymbol{x}_1 \dots \mathrm{d}\boldsymbol{x}_K\ \mathrm{d}\boldsymbol{y}_1 \dots \mathrm{d}\boldsymbol{y}_K\, \delta(e_1)\dots\delta(e_K) \\ &\quad \cdot\delta(f_1)\dots\delta(f_K)p(\boldsymbol{X},\boldsymbol{Y})\log\left(\frac{p(\boldsymbol{X},\boldsymbol{Y})}{p_X(\boldsymbol{X})p_Y(\boldsymbol{Y})}\right). \end{aligned} \tag{D.16}$$

But all the δ functions will integrate to unity, as there is no dependence on e or f. Accordingly we have that

$$I(\boldsymbol{X};\boldsymbol{Y}) = I(\boldsymbol{X}';\boldsymbol{Y}') \tag{D.17}$$

so that our proposition is proved.

E. Example Performance Calculation

Consider a one-dimensional system containing m objects. The measurement constraint indicates that each measurement should be taken every t_c seconds with an error variance of σ_z^2. Each object moves independently of one another. The equation of motion for one object is as follows

$$\frac{dq}{dt} = v + n \tag{E.1}$$

where q is the position, v is a known velocity and n is random noise.

The trajectory of the object, starting from an unknown position at $t = 0$, will be given by

$$q(t) = x_1 + \int_0^t n(t)\,\mathrm{dt} + vt, \tag{E.2}$$

where x_1 is the initial uncertainty in position. This is assumed to be gaussian distributed with zero mean and a variance of σ_0^2.

It is assumed that time can be measured very precisely. Then vt is known accurately, so from the viewpoint of judging information transfer, we can replace $q(t)$ by $q(t) - vt$, yielding

$$x(t) = x_1 + \int_0^t n(t)\,\mathrm{dt}\,. \tag{E.3}$$

The measurement constraint dictates that $x(t)$ should be represented by a sequence $x_1, x_2, \ldots, x_K$ where

$$x_i = x(it_c) = x_1 + \sum_{k=1}^{i} w_k \;;\; i = 2, \ldots, K \tag{E.4}$$

and

$$w_k = \int_{(k-1)t}^{kt} n(t)\,\mathrm{dt}\,. \tag{E.5}$$

Assume that the w_k are independent, zero-mean, gaussian random variables with variance σ_w^2. Furthermore, suppose that analysis indicates that when the system's performance is much less than its capacity the measurement can be represented as follows:

$$y_i = x_i + z_i \;;\; i = 1, \ldots, K. \tag{E.6}$$

where z_i is zero-mean gaussian additive measurement noise. The z_i are independent of the x_i. The measurement constraint dictates that the variance of z_i will be σ^2.

Accordingly, we have that the information transfer is given by

$$I(\boldsymbol{X};\boldsymbol{Y}) = H(\boldsymbol{Y}) - H(\boldsymbol{Z}). \tag{E.7}$$

To evaluate this equation we need to calculate $p_Y(Y)$. Now we have, using (E.4) and (E.6)

$$y_i = \begin{cases} y_{i-1} + w_i - z_{i-1} + z_i & i = 2, \ldots, K \\ x_1 + z_1 & \text{i=1.} \end{cases} \tag{E.8}$$

Assume that each of the w_i are independent, each of the z_i are independent and the z_i are independent of the w_i.

Then y_i can be approximated by a Markov process, provided $\sigma_z^2 << \sigma_w^2$. So, from Papoulis [87, page 529] we have that

$$p_Y(y_1, \ldots, y_K) = p(y_K|y_{K-1}) \ldots p(y_2|y_1)p(y_1). \tag{E.9}$$

The statistics of w_i and z_i, together with (E.8) imply that $p(y_i|y_{i-1})$ will be gaussian with variance $\sigma_w^2 + 2\sigma_z^2$ for $i = 2, \ldots, K$. For $i = 1$ we have that $p(y_1)$ is zero-mean gaussian with variance $\sigma_0^2 + \sigma_z^2$.

Accordingly we can write

$$p_Y(y_1, \ldots, y_K) = \frac{C^{\frac{1}{2}}}{(2\pi)^{\frac{K}{2}}} \exp\left(-\frac{1}{2}\boldsymbol{Y}^{\mathrm{T}}\boldsymbol{A}\boldsymbol{Y}\right) \tag{E.10}$$

where $\boldsymbol{A}$ is a real symmetric banded $K \times K$ matrix and

$$C = \frac{1}{(2\sigma_a^2 + \sigma_w^2)^{K-1}(\sigma_z^2 + \sigma_0^2)}. \tag{E.11}$$

A comparison of (E.10) with Eqn. (6.5.13) of Jones [53] shows that p_Y is, apart from a constant factor, a K-dimensional gaussian distribution. Normalising (E.10) shows that

$$C = \det(\mathbf{A}) \tag{E.12}$$

so using Eqn. (6.5.16) of Jones [53] shows that

$$H(\boldsymbol{Y}) = \log\left(\frac{(2\pi e)^{\frac{K}{2}}}{C^{\frac{1}{2}}}\right). \tag{E.13}$$

Now p_Z will clearly be a K-dimensional probability distribution (joint distribution of K independent zero-mean gaussian variables each with variance σ_z^2). So we have

$$H(\boldsymbol{Z}) = \log\left(\frac{(2\pi e)^{\frac{K}{2}}}{\sigma_z^{2K}}\right). \tag{E.14}$$

Then using (E.7),(E.13) and (E.14) we have that the mutual information transfer for one object for a time period $(K-1)t_c$ is given by

$$I(\boldsymbol{X};\boldsymbol{Y}) = \frac{(K-1)}{2}\log\left(2+\frac{\sigma_w^2}{\sigma_z^2}\right)+\frac{1}{2}\log\left(1+\frac{\sigma_0^2}{\sigma_z^2}\right). \tag{E.15}$$

Accordingly for m independent objects, the source information rate for a time period $T=(K-1)t_c$ is given by

$$S_T = \frac{1}{2t_c}\log\left(2+\frac{\sigma_w^2}{\sigma_z^2}\right)+\frac{1}{2t_c(K-1)}\log\left(1+\frac{\sigma_0^2}{\sigma_z^2}\right). \tag{E.16}$$

In the limit, the source information rate no longer depends on the initial uncertainty in position, so we are left with

$$S = \frac{1}{2t_c}\log\left(2+\frac{\sigma_w^2}{\sigma_z^2}\right). \tag{E.17}$$

F. Classification of Systems

Chapter 2 showed that one of the characteristics of a wave-based system is the loci defining the communications frame. The appendix presents a method of classifying systems based on their loci. The systems described in here make position measurements by measuring two or more loci and calculating the intersection of the loci. In a remote-positioning system, measurements made at the reference sites are conveyed to a central site (or sites) in order to calculate the intersection of the loci. We consider the measurement of locus using four different techniques: angle, range, elliptic and hyperbolic. These are described below:

Angle This measures the direction of arrival of a wave at a reference site. This can be done using a directional antenna or measuring the angle of polarization. An example would be measuring the azimuthal angle of arrival of a plane wave using a linear array. In two dimensions the locus of Angle is a straight line passing through the reference site and in three dimensions it is half plane passing through the reference site.

Radial This measures the distance between the remote object and a reference site. An example would a radar sending out a pulse and measuring the round trip time. In two dimensions the locus is a circle and in three dimensions a sphere. Both are centred on the reference site.

Elliptic This measures the transit time of a wave between a reference site, the object and another reference site. An example would be sending out a narrow pulse from a transmitter, this pulse being received by the object and then being re-transmitted to another receiver which is separate from the object and the first transmitter. In two dimensions the locus would be an ellipse and in three dimensions it would be an ellipsoid. The foci of the ellipse and ellipsoid are the reference sites.

Note that in order to make a transit measurement, the two reference sites must share a common time reference. This can be established by using highly stable oscillators, sending transmissions between the reference sites or broadcasting a time reference to all reference sites. In addition, for a transit measurement to be used in a self-positioning system it is necessary to establish some time reference between the object and the reference sites.

Hyperbolic To derive a hyperbolic locus it is necessary to measure the time difference of arrival of a wave from an object at two reference sites (remote-positioning) or the time difference of arrival of two waves launched from two separate reference sites at the remote object (self-positioning). For example in Loran-C, a shipborne receiver picks up pulses sent from two different Loran stations. The receiver then calculates the time difference of arrival of the two pulses. In two dimensions the locus is a hyperbola and in three dimensions a hyperboloid. The foci of the hyperbola and the hyperboloid are the two reference sites. To make a time difference of arrival measurement the two reference sites must share a common time reference.

Sometimes it will not be immediately obvious if a particular set of measurements from two reference sites should be classified as radial, hyperbolic or elliptic. In this case the following rules should be used:

Rule 1 If one or both of the reference sites and the *object* share a common time reference or the distance between them is known then the locus derived from those two reference sites is *not* hyperbolic.

Rule 2 To be considered a radial measurement, the radial distance must be deducible from measurements between only one reference site and the object.

The reason that these rules are necessary is that both elliptic and hyperbolic systems use pseudo-range loci, they differ only the nature of the *a priori* knowledge concerning the relationship between the pseudo-range loci. For example a hyperbolic system 'knows' that the pulses received at two stations were sent from a single site at the same time.

Using these definitions it is possible to uniquely define a positioning system, based on which locus measurements are made. So a system which measures range and transit time is called a radial-elliptic system.

This classification scheme can be demonstrated by an example. Consider the remote-positioning system shown in Fig. F.1. This systems works in the following manner: Reference site A and reference site B share a common time reference. At $t = 0$, reference site B sends out a pulse. This is received by the object at C. After a known time delay (t_d), the object at C sends out another narrow pulse. This is received at reference site A at time t_A and at reference site B at time t_B. This system measures the range between B and C $\left(\delta_{BC} = \frac{(t_B - t_d)}{2}\right)$ and the transit time between B, C and A ($\delta_{BCA} = t_A - t_d$). Hence it is called a radial-elliptic system. The actual position of the object is determined by calculating the intersection of the radial and elliptic loci (see Fig. F.2). In this case there are two possible intersections. This ambiguity can be resolved using one of the methods described in Sect. 3.3.

In this example the classification rules can be used to exclude other possibilities: it is possible to calculate the range between A and C, using the formula

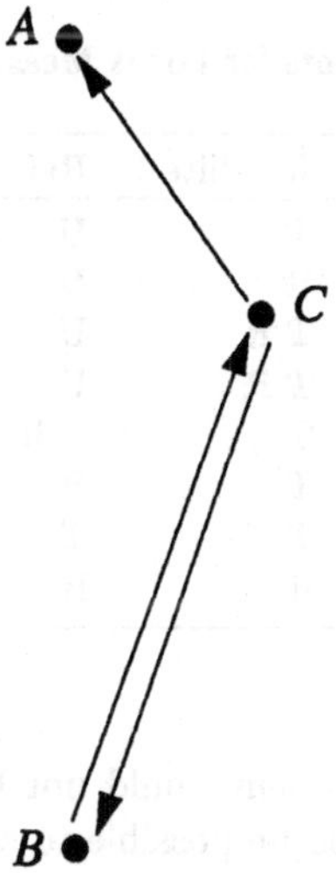

Fig. F.1. Diagram of Example System Configuration

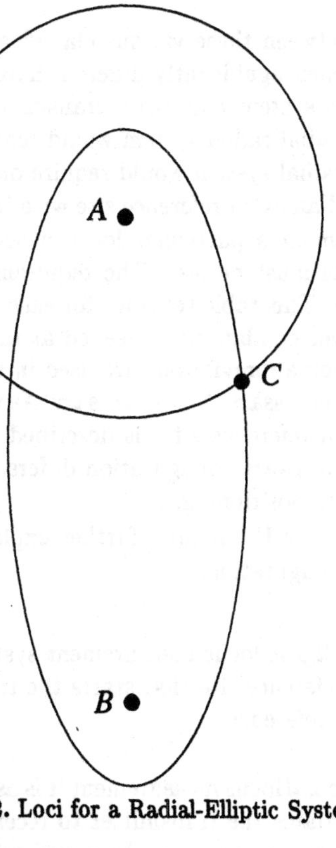

Fig. F.2. Loci for a Radial-Elliptic System

Table F.1. Hardware Requirements for Locus Measurements

Locus	Type	Object	Ref. Site 1	Ref. Site 2	Time Ref.
Angle	Self	R	T	U	U
Angle	Remote	T	R	U	U
Radial	Self	T R	T R	U	U
Radial	Remote	T R	T R	U	U
Elliptic	Self	R	T	T R	N
Elliptic	Remote	T R	T	R	N
Hyperbolic	Self	R	T	T	N
Hyperbolic	Remote	T	R	R	N

$t_{AC} = \delta_{BCA} - \delta_{BC}$. But the system would not be classified as a radial-radial because of Rule 2. It would also be possible to calculate the time difference of arrival at A and C i.e. $t_A - t_B$. However the system would not be classified as radial-hyperbolic because of rule 1, as site B 'knows' when the pulse was sent from C, so establishing a common time reference between the object and site C.

The distinction between these various classifications is of more than theoretical interest, it implies significantly different hardware implementations. For instance for the above system requires a transmitter/receiver at site B and a receiver at site A. A radial-radial system would require an extra transmitter at site A. A hyperbolic-radial system would require only a receiver at sites A and B, but would also need an extra reference site with its own transmitter/receiver.

In order to implement a particular locus measurement, certain minimum hardware requirements must be met. The requirements for each of the loci is described in Table F.1. This table sets out, for each type of locus measurement, the transmitter/receiver combinations needed at the object and the two reference sites. The following abbreviations are used in the table: T = Transmitter, R = Receiver, U = Unnecessary, N = Necessary. As well, the need for a common time reference between reference sites is described in the column titled 'Time Ref.'. Note that the hardware configuration differs, depending on whether the system is self or remote-positioning.

Three entries in Table F.1 require further explanation to define a unique, realizable hardware configuration.

For *self-positioning elliptic* locus measurement systems it is assumed that the object and the reference site that starts the transit pulse share some sort of common time reference.

For *self-positioning radial* locus measurement it is assumed that the transmitter at the object initiates the transmitter to receiver/transmitter to receiver interaction necessary to measure the round trip time at the object.

Table F.2. Two-Dimensional Loci

elliptic	radial
elliptic	hyperbolic
elliptic	Angle
radial	hyperbolic
radial	Angle
hyperbolic	Angle

Table F.3. Three Dimensional Loci

elliptic	elliptic	elliptic
elliptic	elliptic	radial
elliptic	elliptic	hyperbolic
elliptic	elliptic	Angle
elliptic	radial	radial
elliptic	radial	hyperbolic
elliptic	hyperbolic	hyperbolic
elliptic	hyperbolic	Angle
elliptic	Angle	Angle
radial	radial	d
radial	radial	Angle
radial	hyperbolic	hyperbolic
radial	Angle	Angle
hyperbolic	hyperbolic	hyperbolic
hyperbolic	hyperbolic	Angle
hyperbolic	Angle	Angle

For *remote-positioning radial* locus measurement it is assumed that the transmitter at the reference site initiates the transmitter to receiver/transmitter to receiver interaction necessary to measure the round trip time at the reference site.

The above classification scheme together with Table F.1 means that by describing the loci of a system it is possible to define the important hardware aspects of the system. Of course certain combinations of loci are already employed in positioning systems e.g. JTIDs [35] uses a radial-radial system. However many combinations have not been used. Table F.2 lists novel two-dimensional combinations of loci, while Table F.3 sets out three-dimensional loci.

This classification scheme could be extended to inverse positioning systems. An example of an inverse system is a self-positioning system which has a reference site with transmitter with a highly directional antenna. This antenna scans the volume of interest, and places a modulation on the scan so that an object which is momentarily illuminated by the spot beam will know where it is to within the accuracy of the spot.

G. Example of a Class of Systems

The members of the class two-dimensional classes described in Example 4.7 are enumerated below. Only the $G(\alpha)$ is described, because the A matrix can be derived from this information. Representative loci are shown in Fig. G.1. The term radial(r_1) means that radial distance is measured from the first reference site, while radial(r_2) means measured from the second reference site.

1. Radial-radial

$$G = \begin{pmatrix} 2 & 0 \\ 0 & 2 \end{pmatrix} \tag{G.1}$$

2. Radial(r_1)-elliptic

$$G = \begin{pmatrix} 2 & 0 \\ 1 & 1 \end{pmatrix} \tag{G.2}$$

3. Radial(r_1)-elliptic

$$G = \begin{pmatrix} -1 & -1 \\ 2 & 0 \end{pmatrix} \tag{G.3}$$

4. Radial(r_1)-hyperbolic

$$G = \begin{pmatrix} 2 & 0 \\ -1 & 1 \end{pmatrix} \tag{G.4}$$

5. Radial(r_1)-hyperbolic

$$G = \begin{pmatrix} -1 & -1 \\ 2 & 0 \end{pmatrix} \tag{G.5}$$

6. Radial(r_2)-elliptic

$$G = \begin{pmatrix} 1 & 1 \\ 0 & 2 \end{pmatrix} \tag{G.6}$$

7. Radial(r_2)-elliptic

$$G = \begin{pmatrix} 0 & 2 \\ -1 & -1 \end{pmatrix} \tag{G.7}$$

8. Radial(r_2)-hyperbolic

$$G = \begin{pmatrix} 1 & -1 \\ 0 & 2 \end{pmatrix} \tag{G.8}$$

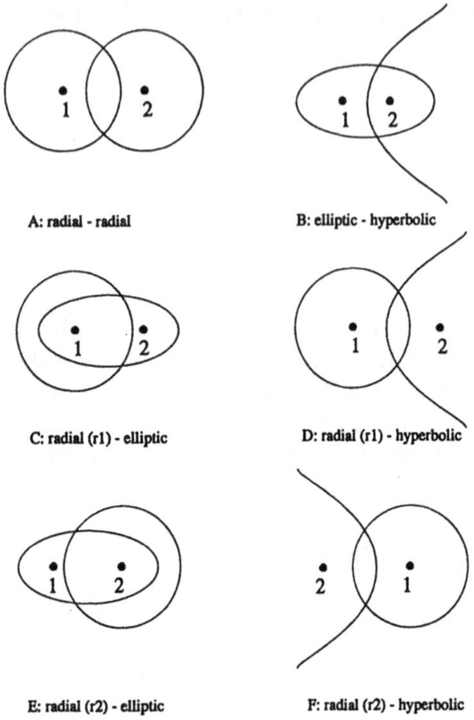

Fig. G.1. Representative Locii

9. Radial(r_2)-hyperbolic

$$G = \begin{pmatrix} 0 & 2 \\ -1 & 1 \end{pmatrix} \tag{G.9}$$

10. Elliptic-Hyperbolic

$$G = \begin{pmatrix} 1 & 1 \\ -1 & 1 \end{pmatrix} \tag{G.10}$$

11. Elliptic-Hyperbolic

$$G = \begin{pmatrix} 1 & -1 \\ 1 & 1 \end{pmatrix} \tag{G.11}$$

12. Elliptic-Hyperbolic

$$G = \begin{pmatrix} -1 & -1 \\ 1 & -1 \end{pmatrix} \tag{G.12}$$

13. Elliptic-Hyperbolic

$$G = \begin{pmatrix} -1 & 1 \\ -1 & -1 \end{pmatrix} \tag{G.13}$$

References

[1]M. Abramowitz and I. A. Stegun. *Handbook of Mathematical Functions.* Dover Publications, New York, 1964.

[2]E. Aslaksen and W. R. Belcher. *Systems Engineering.* Prentice Hall, New York, 1991.

[3]H. Atmanspacher. The aspect of information production in the process of observaton. *Foundations of Physics*, 19(5):553–578, May 1989.

[4]Y. Bar-Shalom and K. Birmiwal. Variable dimension filter for manoeuvering target tracking. *IEE Transactions on Aerospace and Electronic Systems*, AES-18(5):621–629, September 1982.

[5]S. Benedetto, E. Biglieri, and V. Catellani. *Digital Transimission Theory.* Prentice Hall, New Jersey, 1989.

[6]T. Berger. Rate distortion theory for sources with abstract alphabets and memory. *Information and Control*, 13():254–273, September 1968.

[7]T. Berger. Information rates of Wiener processes. *IEEE Transactions on Information Theory*, IT-16():134–139, March 1970.

[8]Berger T. *Rate Distortion Theory, A Mathematical Basis for Data Compression.* Prentice-Hall, Englewood Cliffs, New Jersey, 1971.

[9]S. Bozic. *Digital and Kalman Filters.* Edward Arnold, London, 1979.

[10]R. M Bracewell. *The Fourier Transform and Its Application.* Mc-Graw Hill International Book Company, Sydney, 1978.

[11]Y. Bresler, V. U. Reddy, and T. Kailath. Optimum beamforming for coherent signal and interferences. *IEEE Transactions on Acoustics, Speech, and Signal Processing*, 36(6):833–443, June 1988.

[12]G. Clifford Carter. Coherence and time delay estimation. *Proceedings of the IEEE*, 75(2):236–255, February 1987.

[13]C. Chatterjee, R. L. Kashyap, and B. Boray. Estimation of close sinusoids in colored noise and model discrimination. *IEEE Transactions on Acoustics, Speech, and Signal Processing*, ASSP-35(3):328–337, March 1987.

[14]F. Cheng. *Properties of Toeplitz Matrices and their Application to Vectorial Two-Dimensional Signal Processing.* PhD Thesis, Kyushu University, Japan, 1988.

[15]H. H Chiang and C. L. Nikias. A new method for adaptive time delay estimation for non-gaussian signals. *IEEE Transactions on Acoustics, Speech, and Signal Processing*, 38(2):209–219, February 1990.

[16]L. Cohen. Time-frequency distributions - a review. *Proceeding of the IEEE*, 77(7):941–981, July 1989.

[17]J. P. Costas. A study of a class of detection waveforms having nearly ideal range - doppler ambiguity properties. *Proceedings of the IEEE*, 72(8):996–1008, August 1984.

[18]I. J. Cox and C. J. R. Sheppard. Information capacity and resolution in an optical system. *Journal of the Optical Society of America A*, 3:1152–1158, August 1986.

[19]H. A. d'Assumpcao and G.E Mountford. An overview of signal processing for arrays of receiver. *Journal of Electrical and Electronics Engineering, Australia*, 4(1):9–19, March 1984.

[20]K. Demirbas. Manoeuvering target tracking with hypothesis testing. *IEEE Transactions Aerospace and Electronic Systems*, AES-23(6):757–766, November 1987.

[21]R. A. Dork. Satellite navigation for land vehicles. *IEEE Position Location and Navigation Symposium*, ():414–418, November 1986.

[22]C. R. Drane. An information theoretic approach to characterising radar systems. *RADARCON '90, Adelaide, Australia*, ():453–460, April 1990.

[23]C. R. Drane. An information theoretic approach to characterising radar systems. *RADARCON 90, Adelaide, Australia*, ():453–460, April 1990.

[24]P. J. Duffet-Smith Navigation and Tracking System, *International Patent No. WO 89/01637*, February 1989.

[25]P. J. Duffet-Smith and G. Woan CURSOR - a new Navigation and Tracking System, *24th International Symposium on Automotive Technology*, ():1–7, May 1991.

[26]M. F. Easterling. *Digital Communications with Space Applications*, chapter Applications to Ranging. Prentice-Hall Inc., Englewood Cliffs, N.J, 1964.

[27]R. S. Elliott. *Antenna Theory and Design.* Prentice Hall, Englewood Cliffs, NJ, 1981.

[28]B. D. Elrod, H. A. Bustamante, and F. D. Natali. A GPS receiver design for general aviation navigation. *IEEE Position Location and Navigation Symposium*, ():33–41, December 1980.

[29]R. Fano. *Transmission of Information.* The MIT Press, MIT, Cambridge, Massachusetts, 1961.

[30]P. Fell. Geodetic positioning using a global positioning system of satellites. *IEEE Position Location and Navigation Symposium*, ():42–47, December 1980.

[31]T. Foley. Space operations begin using geostar payload. *Aviation Week and Space Technology*, ():55, July 1988.

[32]R. L. Frank. Current developments in Loran-C. *Proceedings of the IEEE*, 71(10):1127–1139, October 1983.

[33]L. E. Franks. *Signal Theory.* Prentice-Hall International, London, 1969.

[34]A. M. Fraser. Information and entropy in strange attactors. *IEEE Transactions on Information Theory*, 35(2):245–262, March 1989.

[35]W. R. Fried. Principles and simulation of JTIDS relative navigation. *IEEE Transactions*, ():75–84, January 1978.

[36]V. S. Frost. The information content of synthetic aperture radar images of terrain. *IEEE Transactions of Aerospace and Electronic Systems*, AES-19(5):768–774, September 1983.

[37]F. R. Gantmacher. *The Theory of Matrices.* Chelsea Publishing Company, 1959.

[38]A. M. Gerrish and P. M. Schultheiss. Information rates of non-gaussian sources. *IEEE Transactions on Information Theory*, ():265–271, October 1964.

[39]S. W. Golomb. Two-dimensional synchronization patterns for minimum ambiguity. *IEEE Transactions on Information Theory*, IT-28(4):600–604, July 1982.
[40]J. W. Goodman. *Introduction to Fourier Optics*. McGraw Hill, New York, 1968.
[41]I. S. Grashteyn and I. M. Ryzhik. *Table of Integrals, Sums and Products*. Academic Press, New York, 1968.
[42]R. M. Gray. *Entropy and Information Theory*. Springer Verlag, New York, 1990.
[43]L. J. Griffiths and K. M. Buckley. Quiescent pattern control in linearly constrained adaptive arrays. *IEEE Transactions on Acoustics, Speech, and Signal Processing*, ASSP-35(7):917–926, July 1987.
[44]H. F. Harmuth. *Propagation of Non Sinusiodal Electromagnetic waves*. Academic Press, Inc, Orlando, 1986.
[45]F. Harris. Overview of radar imaging. *Radarcon90, Adelaide, Australia*, 1():154–164, April 1990.
[46]J. K Holmes. *Coherent Spread Spectrum Systems*. Wiley, New York, 1982.
[47]H. C. Horing. Comparison of the fixing accuracy of single-station locators and triangulation systems assuming ideal shortwave propagation in the ionosphere. *IEE Proceedings, Pt F*, 137():173–176, June 1990.
[48]G. K. Hurst Quiktrak: a new AVL system developed in Australia. *Proceeding of IREECON '89, Melbourne*, 1():78–80, 1989.
[49]K. Ikeda and K. Matsumoto. Information theoretical characterization of turbulence. *Physical Review Letters*, 62(19):2265–2268, May 1989.
[50]A. H. Jaczwinski. *Stochastic Processes and Filtering Theory*. Academic Press, New York, 1970.
[51]F. Jelinek. Tree encoding of memoryless time-discrete sources with a fidelity criterion. *IEEE Transactions on Information Theory*, IT-15():584–590, September 1969.
[52]M. J. Johnson, M. Jabri, and C. R. Drane. Application of neural networks to manoeuvre detection in Kalman filters (poster). *The first Australian Neural Networks Conference, Sydney, Australia*, ():, 1990.
[53]D. S. Jones. *Elementary Information Theory*. Clarendon Press, Oxford, 1979.
[54]T. Kailath. On multilink and multidimensional channels. *IRE Transactions on Information Theory*, IT-8():260–261, April 1962.
[55]R. E. Kalman. A new approach to linear filtering and prediction problems. *Journal of Basic Engineering*, 82D():35–45, March 1960.
[56]R. E. Kalman and R. S. Bucy. New results in linear filtering and prediction theory. *Journal of Basic Engineering*, 83D():, March 1961.
[57]R. L. Kashyap, S. G. Oh, and R. N. Madan. Robust estimation of sinusoidal signals with colored noise using decentralised processing. *IEEE Transactions on Acoustics, Speech, and Signal Processing*, 38(1):91–1–4, January 1990.
[58]J. H. Kaufman. Information-theoretic study of pattern formation: rate of entropy production of random fractals. *Physical Review A*, 39(3):1420–1428, February 1989.
[59]J. H. Kaufman and G. M. Dimino. Information-theoretic specific heat of fractal patterns. *Physical Review A*, 39(11):6045–6047, June 1989.
[60]M. Kayton. Navigation - ships to space. *IEEE Aerospace and Electronic Systems*, 24(5):474–519, 1988.

[61]R. J. Kelly. Reducing geometric dilution of precision using ridge regression signal processing. *IEEE Plans 88, Position Location and Navigation Symposium*, ():461–469, 1988.

[62]C. H. Knapp and G. Clifford Carter. The generalised correlation method for estimation of time delay. *IEEE Transactions on Acoustics, Speech and Signal Processing*, ASSP-24(4):320–327, August 1976.

[63]W. E. Kock. *Radar, Sonar and Holography, an introduction.* Academic Press, New York, 1973.

[64]V. A. Kotelnikov. *The Theory of Optimum Noise Immunity.* McGraw-Hill Book Company, New York, 1959.

[65]J. J. Kovaly. *Synthetic Aperture Radar.* Artech House, Dedham, MA, 1976.

[66]K. J. G. Kruscha and B. Pompe. Information flow in 1d maps. *Zeitschrift fur Naturforshung*, 43(a):93–104, February 1988.

[67]J. B. Kuipers. Spasyn - an electromagnetic and relative position and orientation tracking system. *IEEE Transactions on Instrumentation and Measurement*, 29(4):462–466, 3 1980.

[68]M. M. Kuritsky and M. S. Goldstein. Inertial navigation. *Proceedings of the IEEE*, 71(10):1156–1176, October 1983.

[69]V. P. Kuthkhov. Estimation of the information sufficiency of radar signals. *Telecommunications and Radio Engineering*, 36:82–84, August 1981.

[70]J. W. Ladd. Three-coordinate positioning within 1 part in 10 million without the gps code. *IEEE Position Location and Navigation Symposium*, ():238–242, November 1986.

[71]D. D. Lee, R. L. Kashyap, and R. N. Madan. Robust decentralised direction-of-arrival estimation in contaminated noise. *IEEE Transactions on Acoustics, Speech, and Signal Processing*, 38(3):496–505, March 1990.

[72]H. B. Lee. Accuracy limitations of hyperbolic multi-lateration systems. *IEEE Transactions on Aerospace and Electronic Systems*, 11(1):16–29, January 1975.

[73]W. C. Y. Lee. *Mobile Cellular Telecommunications Systems.* McGraw-Hill Book Company, New York, 1989.

[74]A. Van Leeuwen, E. Rosen, and L. Carrier. Global positioning system. 1980.

[75]Jack M. Ligon. Differential-gps, a new approach. *IEEE Position Location and Navigation Symposium*, ():22–27, December 1982.

[76]W. Lukosz. Optical systems with resolving powers exceeding the classical limit. *Journal of the Optical Society of America*, 56(11):1463–1472, November 1966.

[77]W. Lukosz. Optical systems with resolving power exceeding the classical limit. *Journal of the Optical Society of America*, 57(7):932–1192, July 1967.

[78]R. Stuart MacKay. *Medical Images and Displays.* John Wiley and Sons, New York, 1984.

[79]A. Macovski. *Medical Imaging Systems.* Prentice-Hall Inc., Englewood Cliffs, New Jersey, 1983.

[80]K. Matsumoto and I. Tsuda. Calculation of information flow rate from mutual information. *Journal of Physics. A:Mathematics General*, 21(6):1405–1414, March 1988.

[81]J. Max. Quantizing for minimum distortion. *IEEE Transactions on Information Theory*, IT-6():7–12, March 1960.

[82]A. A. McKenzie J. A. Pierce and R. H Woodward. *LORAN - Long Range Navigation.* McGraw-Hill Book Company, New York, 1948.

[83]R. J. Milliken and C. J. Zoller. Principles of operation of navstar and system characteristics. *Navigation: Journal of the Institute of Navigation*, 25(2):95–106, Summer 1978.

[84]A. Moghaddamjoo. Eigenstructure variability of the multiple-source multiple-sensor covariance matrix with contaminated gaussian data. *IEEE Transactions on Acoustics, Speech, and Signal Processing*, 32(2):153–167, February 1988.

[85]B. P. Ng. Constraints for linear predictive and minimum-norm methods in bearing estimation. *IEE Proceedings*, 137F(3):187–191, June 1990.

[86]M. A Pallas and G. Jourdain. Active high resolution time delay estimation for larger BT signals. *IEEE Transactions on Signal Processing*, 39(4):781–788, April 1991.

[87]A. Papoulis. *Probability, Random Variables and Stochastic Processes*. McGraw Hill Book Company, New York, 1965.

[88]J. Payton and S. Qureshi. Trellis encoding: what it is and how it effects data transmission. *Data Communications*, ():143–152, May 1985.

[89]P. Z. Peebles. Monopulse radar angle tracking accuracy with sum channel limiting. *IEEE Transactions on Aerospace and Electronic Systems*, AES-21():137–143, January 1985.

[90]F. Pineda. Recurrent backpropagation and the dynamical approach to adaptive neural computation. *Neural Computation*, 1(2):161–172, 1989.

[91]J. T. Pinkston. An application of rate-distortion theory to a converse to the coding theorem. *IEEE Transactions on Information Theory*, IT-15():66–71, January 1969.

[92]J. B. Plant Y. T. Chan and J. R. T. Bottomley A Kalman tracker with a simple input estimator. *IEEE Transactions on Aerospace and Electronic Systems*, AES-18(2):235–240, March 1982.

[93]B. Pompe, J. Kruscha, and R. W. Leven. State predictability and information flow in simple chaotic systems. *Zeitschrift fur Naturforshung.*, 41(a):801–818, June 1986.

[94]B. N. Pshenichny and Yu. M. Danilin. *Numerical Methods in Extremal Problems*. MIR Publishers, Moscow, 1978.

[95]H. R. Raemer. *Statistical Communication Theory and Its Application*. Prentice-Hall, Englewood Cliffs, New Jersey, 1969.

[96]J. M. Richardson and K. A. Marsh Fusion of Multi-Sensor Data, *The International Journal of Robotics Research*, 7(6):78–96, December 1988.

[97]A. W. Rihaczek. *Principle of High-Resolution Radar*. McGraw-Hill Book Company, New York, 1969.

[98]R. J. McEliece *The Theory of Information and Coding*. Addison-Wesley Publishing Company, London, 1977.

[99]W. L. Root. Remarks, mostly historical, on signal detection and signal parameter estimation. *Proceedings of the IEEE*, 75(11):1447–1456, November 1987.

[100]A. P. Sage and J. L. Melsa *Estimation Theory with Applications to Communications and Control*. McGraw-Hill Book Company, New York, 1971.

[101]D. J. Sakrison. The rate of a class of random processes. *IEEE Transactions on Information Theory*, IT-16():10–16, January 1970.

[102]C. E. Shannon. Coding theorems for a discrete source with a fidelity criterion. *IRE. National Convention Record*, 4():142–163, 1959.

[103]C. E. Shannon and W. Weaver. *The Mathematical Theory of Communications*. University of Illinois Press, Urbana, 1949.

[104]R. Shaw. Strange attractors, chaotic behaviour, and information flow. *Zeitschrift fur Naturforshung*, 36(a):80–112, January 1981.

[105]Y. G. Sinai. *Introduction to Ergodic Theory*. Princeton University Press, Princeton, New Jersey, 1976.

[106]R. A. Singer. Estimating optimal tracking filter performance for manned manoeuvering targets. *IEEE Transactions on Aerospace and Electronic Systems*, AES-6(4):567–577, July 1970.

[107]M. I. Skolnik. *Introduction to Radar Systems*. McGraw-Hill Book Company, New York, 1980.

[108]M. I. Skolnik. *Introduction to Radar Systems*. McGraw-Hill Book Company, New York, 1962.

[109]M. I. Skolnik. *Radar Handbook*. McGraw-Hill Book Company, New York, 1970.

[110]D. Slepian. Estimation of signal parameters in the presence of noise. *Transactions of the IRE Professional Group on Information Theory*, PGIT-3():68–89, March 1954.

[111]A.C Smith and G.C.L Searle. A comparison between conventional and adaptive beamformers using underwater data. *International Conference on Developments in Marine Acoustics, Sydney*, 1():175–179, 1984.

[112]S. S. Soliman. Tracking loop for fading dispersive channels. *IEEE Transactions on Communications*, 38(3):292–299, March 1990.

[113]E. R. Swanson. Omega. *Proceedings of the IEEE*, 71(10):1140–1155, October 1983.

[114]H. Taub and D. L. Shilling. *Principles of Communication Systems*. McGraw-Hill, Sydney, 1971.

[115]S. P. Teasley, W. M. Hoover, and C. R. Johnson. Differential gps navigation. *IEEE Position Location and Navigation Symposium*, ():9–16, December 1980.

[116]R. M. Turner and A. W. Bridgewater. An information-theoretic approach to energy management for surveillance. *Sixteenth Asilomar Conference on Circuits, Systems and Computers*, 230–233, November 1982.

[117]G. Ungerboeck. Trellis-coded modulation with redundant signals sets part 1: introduction. *IEEE Communications Magazine*, 25(2):5–21, February 1987.

[118]R. J. Urick. *Principles of Underwater Sound*. McGraw Hill, New York, 1983.

[119]B. D. Van Veen. An analysis of several partially adaptive beamformer designs. *IEEE Transactions on Acoustics, Speech, and Signal Processing*, 32(2):192–203, February 1989.

[120]B. D. Van Veen. Optimisation of quiescent response in partially adaptive beamformers. *IEEE Transactions on Acoustics, Speech, and Signal Processing*, 38(3):471–477, March 1990.

[121]B. D. Van Veen and R. A. Roberts. Partially adaptive beamformer design via output power minimisation. *IEEE Transactions on Acoustics, Speech, and Signal Processing*, ASSP-35(11):1524–1532, November 1987.

[122]A. J. Viterbi. *Digital Communications with Space Applications*, chapter Phase Coherent Communications over the Continuous Gaussian Channel. Prentice-Hall Inc., Englewood Cliffs, N.J, 1964.

[123]H. L. Weidermann and E. B. Stear. Entropy analysis of estimating systems. *IEEE Transaction on Information Theory*, It-16(3), May 1970.

[124]T. J. Willmore. *An Introduction to Differential Geometry.* Oxford University Press, London, 1964.

[125]K. A. Winick. Cramér-rao lower bounds on the performance of charge-coupled-device optical position estimators. *Journal of the Optical Society of America, A*, 3(11):1809 – 1815, November 1986.

[126]J. K. Wolf. Transmission of noisy information to a noisy receiver with minimum distortion. *IEEE Transactions on Information Theory*, IT-16():406–411, July 1970.

[127]P. M. Woodward. Information theory and the design of radar receivers. *Proceedings of the IRE*, 39(12):1521–1524, 1951.

[128]P. M. Woodward. *Probability and Information Theory, with Applications to Radar.* Permagon Press, Oxford, 1964.

[129]P. M. Woodward and I.L.Davies. A theory of radar information. *Philosophical Magazine*, 41(7):1001–1017, October 1950.

[130]A. D. Wyner. Bounds on the rate-distortion function for stationary sources with memory. *IEEE Transactions on Information Theory*, IT-17():508–513, September 1971.

[131]B. Yavorksy and A. Detalaf. *Handbook of Physics.* MIR Publishers, Moscow, 1977.

[132]M. J. Yerbury and G. C. Hurst. Spread spectrum multiplexed transmission system. *Australian Patent, No. 582038*, ():, January 1987.

[133]G. W. Zeoli. A lower bound on the data rate for synthetic aperture radar. *IEEE Transactions on Information Theory*, IT-22(6):708–715, November 1976.

Index

Lecture Notes in Control and Information Sciences

Edited by M. Thoma and A. Wyner

Vol. 137: S. L. Shah, G. Dumont (Eds.)
Adaptive Control Strategies for
Industrial Use
Proceedings of a Workshop
Kananaskis, Canada, 1988
VI, 360 pages. 1989

Vol. 138: D. C. McFarlane, K. Glover
Robust Controller Design Using
Normalized Coprime Factor
Plant Descriptions
X, 206 pages. 1990

Vol. 139: V. Hayward, O. Khatib (Eds.)
Experimental Robotics I
The First International Symposium
Montreal, June 19-21, 1989
XVII, 613 pages. 1990

Vol. 140: Z. Gajic, D. Petkovski,
X. Shen
Singularly Perturbed and Weakly
Coupled Linear Control Systems
A Recursive Approach
VII, 202 pages. 1990

Vol. 141: S. Gutman
Root Clustering in Parameter Space
VIII, 153 pages. 1990

Vol. 142: A. N. Gündeş, C. A. Desoer
Algebraic Theory of Linear
Feedback Systems with Full and
Decentralized Compensators
V, 176 pages. 1990

Vol. 143: H.-J. Sebastian, K. Tammer (Eds.)
System Modelling and Optimization
Proceedings of the 14th IFIP Conference
Leipzig, GDR, July 3-7, 1989
X, 960 pages. 1990

Vol. 144: A. Bensoussan, J. L. Lions (Eds.)
Analysis and Optimization of Systems
Proceedings of the 9th International
Conference, Antibes, June 12-15, 1990
XII, 992 pages. 1990

Vol. 145: M. B. Subrahmanyam
Optimal Control with a Worst-Case
Performance Criterion and Applications
X, 133 pages. 1990

Vol. 146: D. Mustafa, K. Glover
Minimum Entropy H_∞ Control
IX, 144 pages. 1990

Vol. 147: J. P. Zolésio (Ed.)
Stabilization of Flexible Structures
Third Working Conference,
Montpellier, France, January 1989
V, 327 pages, 1990

Vol. 148: In preparation

Vol. 149: K.H. Hoffmann, W. Krabs (Eds.)
Optimal Control of Partial
Differential Equations
Proceedings of the IFIP WG 7.2 International
Conference, Irsee, April 9-12, 1990
VI, 245 pages. 1991

Vol. 150: L. C. G. J. M. Habets
Robust Stabilization
in the Gap-topology
IX, 126 pages. 1991

Vol. 151: J. M. Skowronski,
H. Flashner, R. S. Guttalu (Eds.)
Mechanics and Control
Proceedings of the 3rd Workshop on
Control Mechanics
in Honor of the 65th Birthday of
George Leitmann
January 22-24, 1990, University of Southern
California
IV, 497 pages. 1991

Vol. 152: J. D. Aplevich
Implicit Linear Systems
XI, 176 pages. 1991

Vol. 153: O. Hájek
Control Theory in the Plane
X, 272 pages. 1991

Vol. 154: A. Kurzhanski, I. Lasiecka (Eds.)
Modelling and Inverse Problems of Control
for Distributed Parameter Systems
Proceedings of IFIP (W.G. 7.2)-IIASA Conference
Laxenburg, Austria, July 24-28, 1989
VII, 179 pages. 1991

Vol. 155: M. Bouvet, G. Bienvenu (Eds.)
High-Resolution Methods in Underwater
Acoustics
V, 249 pages. 1991

Lecture Notes in Control and Information Sciences

Edited by M. Thoma and A. Wyner

Vol. 156: R. P. Hämäläinen, H. K. Ehtamo (Eds.)
Differential Games –
Developments in Modelling and Computation
Proceedings of the Fourth International Symposium on Differential Games and Applications
August 9-10, 1990, Helsinki University of Technology, Finland
XIII, 292 pages. 1991

Vol. 157: R. P. Hämäläinen, H. K. Ehtamo (Eds.)
Dynamic Games in Economic Analysis
Proceedings of the Fourth International Symposium on Differential Games and Applications
August 9-10, 1990, Helsinki University of Technology, Finland
XIII, 311 pages. 1991

Vol. 158: K. Warwick, M. Kárný, A. Halousková (Eds.)
Advanced Methods in Adaptive Control for Industrial Applications
X, 331 pages. 1991

Vol. 159: X. Li, J. Yong (Eds.)
Control Theory of Distributed Parameter Systems and Applications
Proceedings of the IFIP WG 7.2 Working Conference, Shanghai, China, May 6–9, 1990
VIII, 219 pages. 1991

Vol. 160: P. V. Kokotović (Ed.)
Foundations of Adaptive Control
IX, 525 pages. 1991

Vol. 161: L. Gerencsér, P. E. Caines (Eds.)
Topics in Stochastic Systems:
Modelling, Estimation and Adaptive Control
IV, 401 pages, 1991

Vol. 162: C. Canudas de Wit (Ed.)
Advanced Robot Control
IX, 314 pages, 1991

Vol. 163: V. L. Mehrmann
The Autonomous Linear Quadratic Control Problem
Theory and Numerical Solution
VI, 177 pages, 1991

Vol. 164: I. Lasiecka, R. Triggiani
Differential and Algebraic Riccati Equations with Application to Boundary/Point Control Problems: Continuous Theory and Approximation Theory
XI, 160 pages, 1991

Vol. 165: G. Jacob, F. Lamnabhi-Lagarrigue (Eds.)
Algebraic Computing in Control
Proceedings of the First European Conference
Paris, March 13–15, 1991
IX, 385 pages. 1991

Vol. 166: L. L. M. van der Wegen
Local Disturbance Decoupling with Stability for Nonlinear Systems
V, 135 pages, 1991

Vol. 167: M. Rao
Integrated System for Intelligent Control
VIII, 133 pages. 1992

Vol. 168: P. Dorato, L. Fortuna, G. Muscato
Robust Control for Unstructured Perturbations – An Introduction
VI, 117 pages. 1992

Vol. 169: V. M. Kuntzevich, M. Lychak
Guaranteed Estimates, Adaptation and Robustness in Control Systems
IV, 209 pages. 1992

Vol. 170: J. M. Skowronski, H. Flashner, R. S. Guttalu (Eds.)
Mechanics and Control
Proceedings of the 4th Workshop on Control Mechanics, January 21–23, 1991
University of Southern California, USA
IV, 301 pages. 1992

Vol. 171: P. Stefanidis, A. P. Papliński, M. J. Gibbard
Numerical Operations with Polynomial Matrices
Application to Multi-Variable Dynamic Compensator Design
VIII, 205 pages. 1992

Vol. 172: H. Tolle, E. Ersü
Neurocontrol
Learning Control Systems
Inspired by Neuronal Architectures and Human Problem Solving Strategies
X, 211 pages. 1992

Vol. 173: W. Krabs
On Moment Theory and Controllability of One-Dimensional Vibrating Systems and Heating Processes
VII, 174 pages. 1992